Architectural Material 2

지은이	담디 편집부 엮음
펴낸이	서경원
편집	나진연
디자인	이철주
펴낸곳	도서출판 담디
등록일	2002년 9월 16일
등록번호	제 9-00102호
주소	서울시 강북구 삼각산로 79, 2층
전화	02-900-0652
팩스	02-900-0657
이메일	damdi_book@naver.com
홈페이지	www.damdi.co.kr

Compiler	DAMDI Publishing House
Publisher	Kyongwon Suh
Editor	Jinyoun Na
Art Director	Cheolju Lee
Publishing	DAMDI Publishing House
Office Address	2F, 79, Samgaksan-ro, Gangbuk-gu, Seoul, 01036, Korea
Tel	+82-2-900-0652
Fax	+82-2-900-0657
E-mail	damdi_book@naver.com
Homepage	www.damdi.co.kr

정가 88,000원
Printed in Korea
ISBN 978-89-6801-081-1 (94540)
978-89-6801-079-8 (set)

이 도서의 국립중앙도서관 출판예정도서목록(CIP)은 서지정보유통지원시스템 홈페이지(http://seoji.nl.go.kr)와 국가자료공동목록시스템(http://www.nl.go.kr/kolisnet)에서 이용하실 수 있습니다.(CIP제어번호: CIP2018024917)

Architectural Material 2

WOOD

담디
DAMDI

Contents

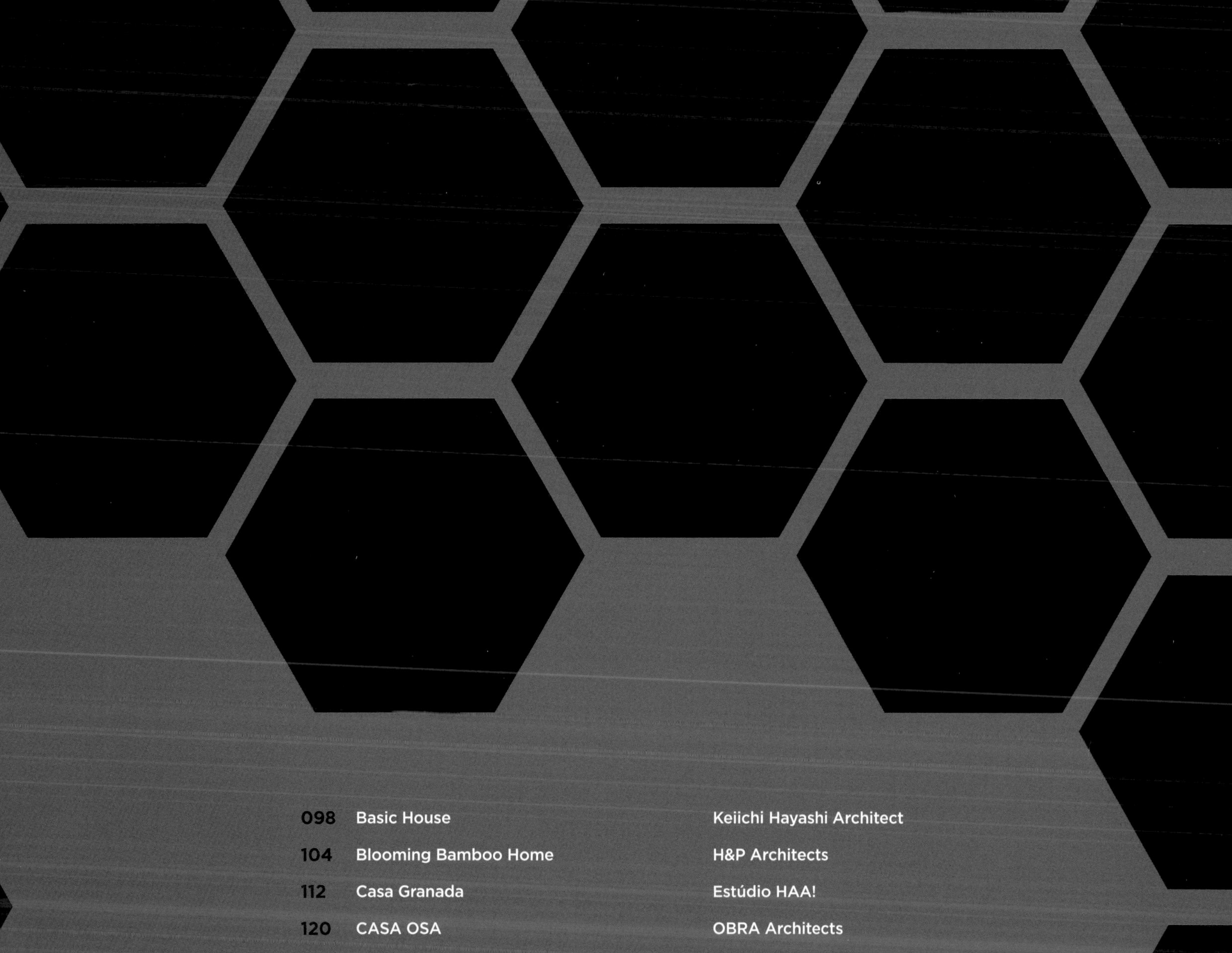

Multi - Housing

Education

Commercial

Office

Installation

Public (Urban & Landscape)

HOW use the WOOD

Interview with Architects

Q1

Tell us about your favourite project that you used Wood in or another architect's work

- it can be in the interior, on the facade, doesn't matter where it's used.

Arenas Basabe Palacios Arquitectos

We have explored the potential of wood to become an element of reference in the project for the renovation of a former industrial warehouse and its conversion into a center for a cultural association. Our 'Box in the box(**314p**)' accommodates the totality of servant spaces (circulation, foyer, toilets, storage and MEP services) within a massive wooden core made of OSB boards. The engineered lumber boards clad every surface of the core: floors, walls and ceilings. The employment of wood confers a recognizable character to this new component, boosting navigability and readability by users from both indoors and outdoors, as well as further consolidating the identity of the building as a cultural center.

ARPHENOTYPE

Well, since this is a book about materials and probably they are countless buildings on the planet, many of them made of wood, I would like to compare two projects, which are from different times, but the same technique. On the one hand, there is the Naiju Residential Center and Kindergarten in Chikuho, Fukuoka, by Shoei Yoh (1995), who uses a special grass: bamboo. On the other side is the Center Pompidou(***A1***) in Metz (2010) by Shigeru Ban and Jean de Gastines with a timber-beam structure. Both projects are a bamboo / wood matrix that creates a freeform that is clad with another material. The differ in the use of computers. In 1995, there weren't really computers or software's available, which could compute such a complexity. I am not sure, but I guess that they used models, like Frei Otto (Multihalle Mannheim, 1975) in its time. Shigeru Ban had the possibility to collaborate with the Swiss/German company Designtoprocution, which generated an algorithm, which directly communicated with the machines. With this algorithm it was possible to CNC mill 18.000 meters of the timber-structure and to produce about 1,800 double-curved glulam segments. So we see that certain experimental wood constructions have been possible for decades, but it becomes feasible through the use of algorithms, which reflect the idea of mass-customization. Also we have to admit, that bamboo and wood are two materials with different identities. Bamboo itself has a history in Asia, but not really in Europe or America, but it should be rethought, as it is a perfect building material. Perhaps it would also be ideal to renature opencast mining areas in Europe. It could be a unique business model that replaces coal with fast growing green bamboo.

AZC

The "Gymnase" is a sports hall in Strasbourg(***A2***, **044p**). We massively used wood in this project.

BOARD

Not so long ago we worked on a design for a tourist centre - a recreational ski complex with 27 ski slopes in the area of Klekovaca Mountain in Bosnia and Herzegovina. When we realized during the design process that for the creation of the ski slopes a large number of trees had to be removed, we decided to re-use them. Thus, we proposed to use the wood from the trees to build the new tourist centre. In that way the wood could be used not only for the structures, but also for

A1 Center Pompidou ©Dietmar Köring

A2 "Gymnase" is a sports hall in Strasbourg

A3 Fisher house ©Jerrye & Roy Klotz, MD

the facades of most of the buildings. Accordingly, we called the project "Out of the Woods". With the help of the trees and by using the different geographical features of the site, we suggested to organize the tourist centre in 3 different areas with 3 different layouts and atmospheres embracing the natural beauty of each location: a compact Main Resort Centre next to the skiing plateau and the ski slopes; a Hotel and Sports Complex on the nearly flat area of the site with individual buildings standing free as the trees in the context; and a Climatic Health Resort Zone in the pine woods.

Carlos Lampreia

In Fisher house(***A3***), a wood house made by Louis Khan, i found amazing the way wood creates a corner space at the living room, joining the external cover with a window, a bench, and some furniture pieces, all made by the same type of wood in the same piece of design.

A4 THE NEST

Casanova+Hernandez Architects

At this moment our office is developing a museum on an existing platform on the waters of a lake in Shkodra, Albania. This museum(***A4***) is part of a bigger development, also developed by our office, in which the intervention area covers an extension of 3,7 km of the shore of Shkodra Lake and 5 km of a new pedestrian and bicycle path in the mountains that runs parallel to the shore. Both together, the renewed waterfront and the new mountain route create a circular pedestrian and bicycle route of 10 km that offers locals and visitors alike a wide variety of recreative, cultural and tourist facilities. These facilities have been divided into 21 architectural and landscape interventions, which are connected to 22 museographic interventions. One of those architectural interventions is this museum made mainly of wood.

The selection of wood for the facades of the museum allows us to integrate it better in its natural environment. The facades are made with charred Western Red Cedar planks and wooden bars of untreated Siberian Larch on top of the planks. The charring process forms a black, sober appearance on the planks, and the untreated wooden bars on top of them create contrast in color and texture. This will create a vibrant facade in neutral colors that will change over time. The facade will reflect the passage of time and will get more integrated into the colors of the surroundings.

A5 Swisshouse XXXII ©Alexandre Zweiger - Lugano TI

CEBRA(Mikkel Frost)

There are many fantastic wooden buildings in the world – both new and old. Lately, I have been particularly fascinated by the wooden structures designed by Kengo Kuma. The structural qualities and the filigree patterns he makes are just amazing.

I am also quite happy with our own Sustainable Club House project. It's small and bit overlooked but as it happens it became "the mother" of many bigger CEBRA projects such as the Smart School in Irkutsk. Here structure and cladding become one in a vertical symbiosis that is quite successful.

Davide Macullo Architects

We have just built a house "Swisshouse XXXII(***A5***, **218p**)" in the Swiss mountains together with the artist Daniel Buren. The wood structure recalls the verticality of the forests and we used local wood. The archetype of the house has been developed one step further and the collaboration with the artists makes this building a sculpture and a house at same time. It is a sculpture to live within. For hundreds of years we have discussed about integration between art and architecture and this is one of the rare examples where the two arts really cooperate. Without the art the building doesn't exist because the art is a structural part of the building. As we have the walls designed in collaboration with Daniel Buren, the beams of the roof are designed in collaboration with another conceptual artist from the region, Miki Tallone.

Donner Sorcinelli Architecture

Cross laminated timber panels have been used for CD House's structure in order to achieve high thermal insulation and environmental friendly design. Time and costs of construction have been reduced and kept under control.

A6 Kinkaku-ji ©Martin Falbisoner(Left), A7 Gassho

Katsutoshi Sasaki+Associates

Japanese classical temple(***A6***).

Keiichi Hayashi Architect

Gassho(***A7***) / Koji Kakiuchi

It is a small shelter created on the remnants of a concrete foundation of a home that was swept away by the 2011 Tsunami in northern Japan. It was entirely made of wood and designed in a single day of DIY construction. I wish to pay my greatest respects to this architect who bravely took action and creatively utilized characteristics of wood in this design.

LANDÍNEZ+REY arquitectos

One of the lesser-known episodes of Le Corbusier's work(***A8***) refers to the set of buildings that the architect would make for his personal vacations in Roquebrune-Cap Martín, his place of retirement in the French Riviera.

A8 Le Corbusier's Cabanon ©Tangopaso

modostudio

The 17th century wooden Farnese theater(***A9***) in Parma is an incredible masterpiece. It was built as a private theater for Farnese family and it is an example of detailed artcrafts.

Mork-Ulnes Architects

Sverre Fehn's Villa Schreiner(***A10***) use of wood both as bearing structure and as finish material is a fantastic project in the way it relates to the site, manipulates light, creates texture and life in a building.

murmuro

We have used wood in a project of an itinerant art pavilion for the Serralves Foundation. The structure, walls, windows, doors and furniture were all in wood, allowing us to assemble / disassemble the pavilion in a simple and fast manner. With this material we were able do design complex connection joints that allowed us to reduce the disassembled building to single elements; a fast, easy and light solucion in terms of transportation that also means less storage area required when in between exhibitions.

A9 Parma-Farnese Theater ©karaian

A10 Villa Schreiner ©Vidariv

NISHIZAWA ARCHITECTS

About our favourite project that use wood in is the Agri Chapel from Yo Momoeda Architects - a Japanese-wooden chapel with a fractal structure system

We love this one because the way they combine the Japanese traditional wood system with the gothic style in contemporary way creatively. And another reason is the way connect the activity of the chapel to the natural surroundings seamlessly.

OOIIO Architecture

I did not work with wood as main material for a building. In our dry and too sunny weather in Spain, wood is not a material very common in architecture, the sun can destroy it if you use it in exteriors. Wood has been used traditionally only in structure for ancient constructions.

I have work with wood in our refurbishment projects where the slabs or roof structures are made with this material, so my experience with wood is more related to building rehabilitation and how to treat the wood to recover its mechanical properties.

A11 TOCOMADERA retail store

Piuarch

For the redevelopment project of the old building Latteria Sociale Valtellina, we decided to use a material widely available in mountain areas: wood. We created one single roof overhanging at the front and side, and we used the wood in the opaque cladding of the façade, in the structure, and as the interior covering of the roof.

SLOT STUDIO

One of our most recent projects is TOCOMADERA retail store(***A11***), which literally means touch the wood. It is an interior design project where we had to find the correct expression and balance of the woods that are in display, with different finishes and varnishes on the floor, walls and furniture. It is a project that exudes warmth and that sets wood as an essential element in the handcrafted sense that tries to evoke the products sold by the store.

A12 Sou Fujimoto's Wooden House ©Kenta Mabuchi

Stefano Corbo STUDIO

Sou Fujimoto's Wooden House (2006, ***A12***) is an example of innovative use of wood in architecture. In this project, in fact, there is no separation between floors, walls and ceiling. Its formal genesis derives from the irregular stacking of 350 mm square profile cedar beams.

The end result is a perforated wooden box, which can trigger unusual spatial articulations

stpmj Architecture

Shear House /stpmj, the project of wood frame structure with wood finish. Sliced and shifted gable roof in one monolithic wood material.

SUPA architects schweitzer song

Wood to us is endlessly fascinating. It is the most fundamental material in architecture. Both of us were socialized by wood. Ryul was raised in a Korean Hanok, I was raised in an Upper-Austrian farmhouse. This experience still dominates our thinking, our aesthetics, and our architectural strategies. Although both houses seem to be physically so far apart, philosophically they provided a common ground between our cultures for us to work with.

TAKK Architecture

The Wooden House(***A12, Front Page***) by Sou Fujimoto. By just using one material in a certain shape (wooden prisms) and aggregating them in a certain way, he manages to create a very rich space.

TOUCH Architect

It is our project names, Chan'4 ECO-STAY(***A13,14***), which is located at Chantaburi, Thailand. It derived from characteristics of the site itself. Chantaburi province contains various types of geography, especially, sea and hill are the main typography of this location. This site does not only have sea and hill, but also has a swamp/ lake and mangrove inside. This homestay contains both accommodation and an organic farming, which is the owner's activity doing agriculture after retirement. So, this farm becomes a part of people who stay here that everyone can live and learn with ecology.

There are three levels which integrated the homestay within the house. Main approach is leading to lobby space at the ground floor that keeps characteristics of Thai traditional house which has an open space with flow circulation on the first floor, called "Tai-Toon". On the second floor, each rooms are suitable for family which can be sighted to the lake and sea. Moreover, there is common space on the roof top for multi-purpose activities for all.

Material using of the overall building is wood. It has been used from floors, walls, roof structure, to all façade elements, except the primary structure, column and beam, which is made of reinforced concrete. The reason is, using all wood in stead of steel can help avoiding extremely rust which occurs because of the sea salt. In addition, in terms of beauty and function, an adjustable wooden trellis façade allows for clear view and natural ventilation and harmonize with nature.

UNStudio

We used wood in the facade cladding of a single family house that we completed a few years ago, called the W.I.N.D. House(***A15,16, 036p***). The wood is hydro-thermic treated, which results in a long durability, while the wood still appears untreated and natural. In time, it changes to a grey colour, which refers to the tones of driftwood found on a nearby beach and the dunes that surround the house.

A13 Chan'4 ECO-STAY

A14 Chan'4 ECO-STAY

A15 W.I.N.D. House Facade ©Fedde de Weert

A16 W.I.N.D. House ©Fedde de Weert

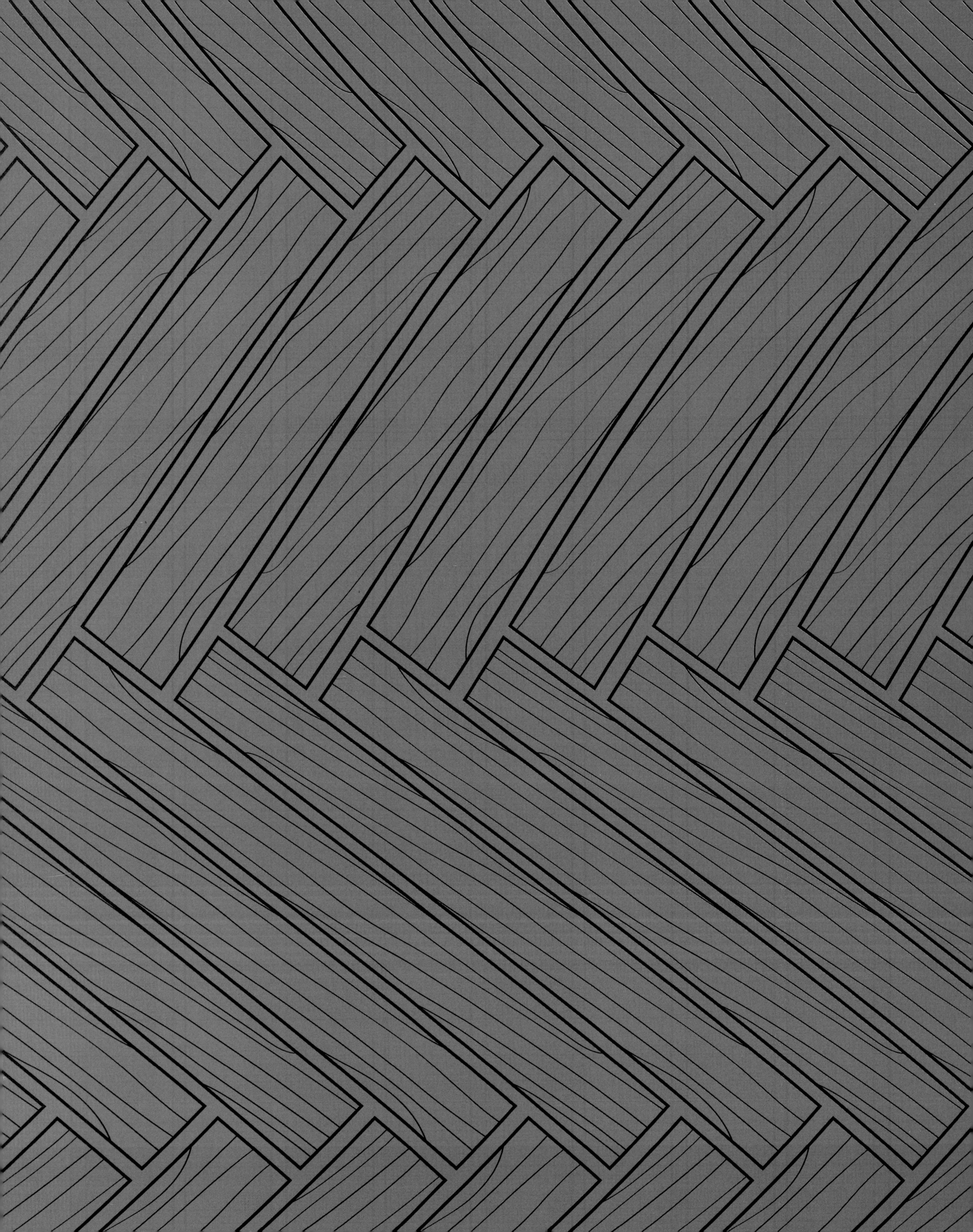

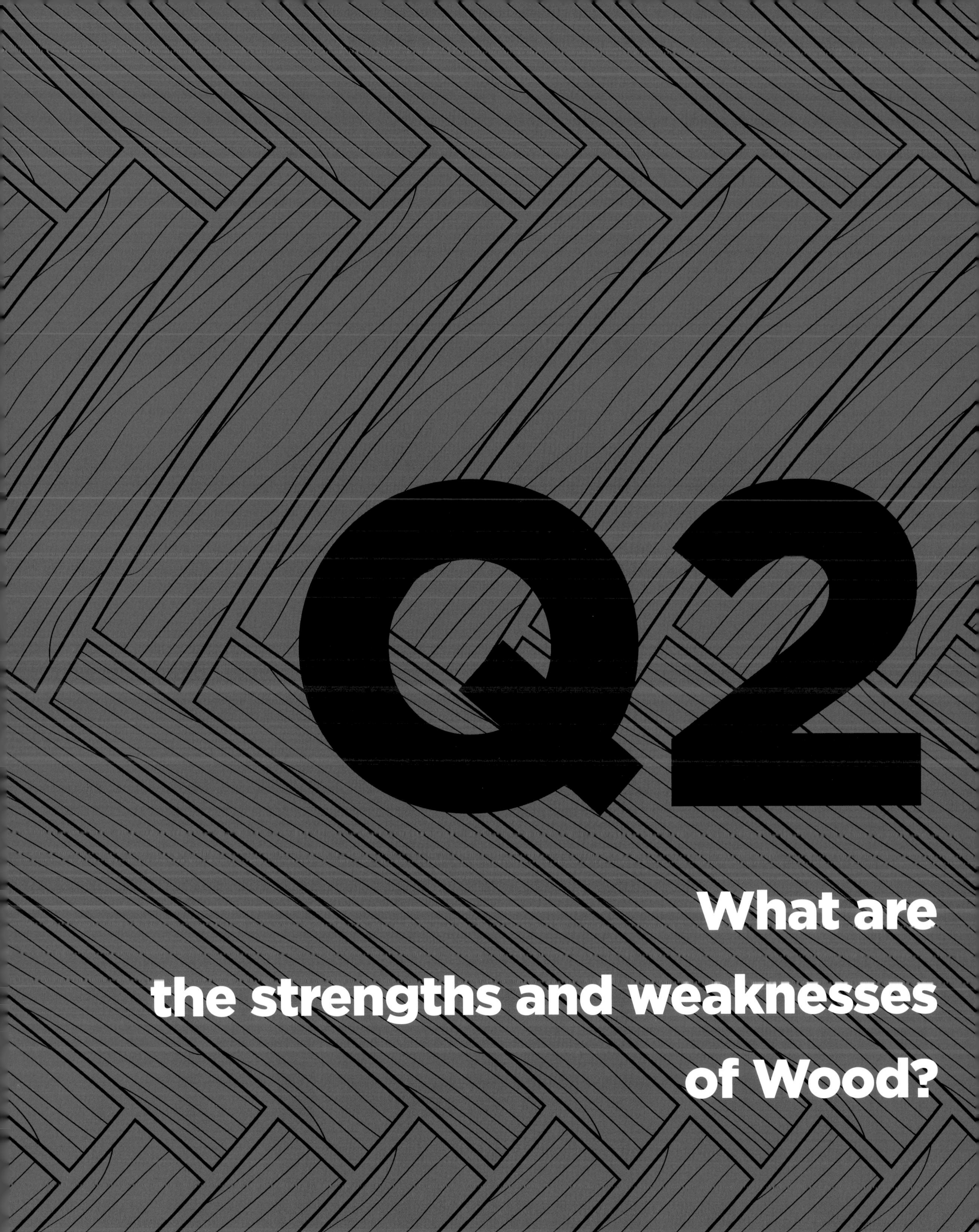

Q2

What are the strengths and weaknesses of Wood?

Arenas Basabe Palacios Arquitectos

The presence of wood in a project usually entails the existence of a strong duality between it and the rest of the architectural components. In our 'Box in the box' project, the forcefulness of the OSB boards that clad the servant core greatly contrasts with the clean lines and neutral-toned materials (grey continuous polished-concrete floors and a mostly white palette of translucent, transparent and opaque walls) employed in the main rooms: indoor playing court, assembly hall, chapel, classrooms, meeting room and offices. The idea of duality can also be acknowledged between the inner wooden box and the new metallic shell that wraps the existing facade, making the idea of a 'Box in the box' more meaningful.

ARPHENOTYPE

Wood is replenished daily by nature in large quantities. With sustainable forestry and responsible consumption, sufficient wood is available for all building purposes at all times. Wood, when properly processed and used purposefully, offers a durability that far exceeds that of most other products. Some wooden buildings of the last 200 to 500 years are often still standing. However, wood is flammable and can be attacked by pests, bacteria or insects. Bamboo impressed above all by its enormous growth speed. Bamboo is resistant and durable even under heavy use. It is very light and elastic thanks to the cavities in the cell structure.

AZC

Is an ecological material, very warm and human, with a low CO2 impact.

It does need protection when used on the exterior. It does need fire protection when used as a main structure.

BOARD

There are a large number of advantages that made wood one of the most favourite construction materials of humans for thousands of years. Some of its strengths are without doubt related to its abundant availability, the ability to renew the supply, but also - and in particular - to its longitudinal shape, its toughness but at the same time lightness, and its easy formability and the ease of making joints due to the softness of most of the trees, which makes it a great material to build all kinds of things, such as houses, boats, furniture, to name just a few. However, one of the weaknesses of wood that I would like to point out is that it can not be produced at a particular location in the way bricks, tiles, or glass can be fabricated - trees need to grow for a long time before the wood can be used – and harvest and transport come with a lot of challenges. In the "Out of the Woods" project we wanted to avoid the difficulties related to the transport of wood by using it right at the spot for the construction of the new tourist centre.

Carlos Lampreia

Wood and earth where the first materials mankind used to provide shelter. With wood we can build almost everything from an house to furniture. In this sense having wood its like have a friend at home, since its a very organic and lived material that moves, sounds and smell, we may even look at it as an artificial tree. The only problem lies in the fact that it is impossible to have an entire city build on wood because we also need forests. This indenible reality makes wood an expansive an rare material that we should use with care.

Casanova+Hernandez Architects

The use of wood with the FSC mark is eco-friendly and help us to minimize the carbon footprint of the building in its natural environment. Its color and texture, which will change with time following a natural process, will remind visitors the nests of the birds in the area and its smell will awake poetic resonances of the woods that were present in the past in the mountains bordering the lake.

On the one hand, wood is a material that provides comfort and that is loaded with poetic meanings, capable of evoking nature and of provoking healing experiences thanks to its visual properties, its touch, and its profound smell.

On the other hand, wood, being an eco-friendly material, helps designers to create more sustainable environments, with respect for nature.

CEBRA(Mikkel Frost)

Fire regulations have been a challenge for many years – if one builds many floors constructed by wood. Many clients also complain about maintenance saying that wood needs constant care, but I disagree with that. In many ways wood is a fantastic material. You can use it for almost anything. It´s cheap and easy to work with and in terms of structure and statics it performs really well. The funny thing is that it just grows out there in the forest. You simply need to cut it down and it´s almost ready for use – no baking, no mixing or reinforcing. If you need more, you simply plant it and wait. In the meantime, it´s a pleasure to watch and a home for birds. It even produces oxygen for the planet!

Wood is not ideal in all situations – as stated above, no material is – but it comes pretty close. I´ve just had my whole yard covered with hard wood. It´s great to walk on it barefoot and it heats up nicely in the sunlight.

Davide Macullo Architects

Wood is one of the most beautiful materials because it is the closest to nature and Man in its use.

Donner Sorcinelli Architecture

Wood represents the sum of all qualities we use to require to a material. It is strong enough to carry loads, it is natural, recyclable as well as warm by a visual and tactile point of view.

Katsutoshi Sasaki+Associates

Strength: It is soft.

Weakness: It is soft.

Keiichi Hayashi Architect

Strength: It is easy for all aspect and has good touch feeling.

Weakness: It burns easily.

LANDÍNEZ+REY arquitectos

We must use the materials of our environment: those that the earth offers us in its proximity. We do not find any weakness in the use of wood in architecture. Any weakness in its use is purely cultural or linked to the absence of this material.

modostudio

The wood is a material that can be used in various situation. It can be a structural construction material, it can be used as a decorative material. Wood is very flexible and has a strong link to the nature.

Mork-Ulnes Architects

Wood can be economical, sustainable, have a unique appeal to the human senses (tactility, olfactory, visual) and incredibly versatile - its only real limitation is in extremely large structures, when in direct contact with water or under certain fire constraints.

murmuro

Wood is an incredible material, it can be crafted to almost every detail you can imagine. It vives a sense of warmth and comfort , it connects with people.The biggest weakness of wood is its maintenance as it requires constant care so it doesn't degrade. If a wood building is well thought and maintained, it can last for centuries.

NISHIZAWA ARCHITECTS

About the strong point of the wood is that it can be used as structure for the whole building without any concrete. Beside that, the color and texture of natural wood are also the elements that help to show the emotion of the building itself also.

About the weak points, wood for us is also a sensitive material that's difficult to use in different environments so it'll be damaged easily by the temperature, humid, insect infestation. The second is that the structural load is also depended on what kind of wood we use, therefore this structure have to combine with some steel elements for increasing bearing capacity sometimes.

OOIIO Architecture

Strength: great structural behavior, is a recyclable material and its production can be sustainable if you work with wood from sustainable forests (FSC forests). Is the only material that produces oxygen when is been produced (growing).

Weakness: Can be affected by parasites, fire, and is not the best material for dry and hot weathers like the one I use to work in.

SLOT STUDIO

Wood is a very special element that has nobleness, warmth and a sense of craftsmanship. We like to use wood in our projects because it has character and does not require any processes to reveal its charm. Wood can transmit a very wide range of sensations just by playing with its dimensions, malleability, textures and its tonalities; furthermore is a sustainable and acoustic material.

Stefano Corbo STUDIO

Defining pros and cons in the use of wood is difficult, as every material should be considered in relation to the overall logic of the project, and based on the relationship between its different components.

stpmj Architecture

Strength: Natural, light, easy to install. Environmental-friendly-material that stimulate touch, smell, and as well as sight simultaneously.

Weakness: The consideration of contraction and expansion of the material and the continuous maintenance are required.

SUPA architects schweitzer song

The overwhelming strength of wood as a building material is its metaphysical quality. Wood lives and reflects our own lives when living within it. It ages, it moves, it makes sounds, it warms, it scents, it changes.

TAKK Architecture

Wood is a material that we use a lot, both in structures and façades, it is cheap and ecological, and has very good thermal properties.

TOUCH Architect

It cannot be denied that every material has their own weaknesses, wood either. Only hard wood can be used for structure which is much more expensive and hard to find, while soft wood can only be used for interior finishing, furniture, and cladding, which is not termite resistance. Some of them are easily to stretch and retract which is not suitable for using as a door or window frame.

Apart from weakness, tons of wood's strengths are already mentioned above, it is physically strong, which is suitable for being a primary structure of the building, while also light and flexible compares to other structure such as concrete and steel. It is yet sustainable and environmental friendly construction material which is the only renewable one. Moreover, wood is not being used for only structure, it is also used for finishing which convey harmonious feeling to the nature. It looks soft and comfy with style.

UNStudio

People often worry that wood will not last long enough, or will require too much maintenance, but in fact today there are treatments that can make wood very durable.

Q1
본인이 목재를 사용한 프로젝트 중 가장 마음에 드는 작품 또는 다른 건축가의 작품을 소개해 달라. (인테리어, 파사드 등 어디에 사용했든 관계없다.)

Arenas Basabe Palacios Arquitectos

우리는 전 산업 창고의 개조 및 문화 협회 센터로 전환하는 프로젝트에서 목재의 잠재력이 판단 요소가 될 가능성을 탐구해보았다. 우리의 '상자 속의 상자' 프로젝트는 OSB 보드로 만든 거대한 나무 코어 안에 봉사하는 공간(동선, 휴게실, 화장실, 저장 및 MEP 서비스)을 모두 수용한다. 공학목재로 바닥, 벽 및 천장과 같은 코어의 모든 표면을 덮었다. 목재로 이 새로운 구성에 알아보기 쉬운 특성을 부여하고, 실내 및 실외 사용자의 가항성과 가독성을 높이고, 문화 중심지로서의 건물의 정체성을 통합했다.

ARPHENOTYPE

이 책은 재료에 관한 책이고, 아마 지구상에는 수많은 건물이 있으며, 그중 많은 수가 나무로 만들어졌을 것이기 때문에 나는 같은 기술을 쓰지만 다른 시대의 프로젝트 두 개를 비교하고 싶다. 하나는 요우 쇼헤이 (1995년)가 후쿠오카 치쿠호에 지은 나이주 주거 센터 및 유치원이다. 그는 이 건물에 특별한 식물을 썼는데, 이는 바로 대나무이다. 다른 하나는 반 시게루(Shigeru Ban)와 장 드 가스틴(Jean de Gastines)이 메츠(2010년)에 목재 들보 구조로 디자인한 퐁피두 센터이다. 두 프로젝트 모두 다른 재료로 덮고 자유형을 만드는 대나무/나무 매트릭스이다. 차이가 있다면 컴퓨터이다. 1995년에는 그런 복잡성을 계산할 수 있는 컴퓨터나 소프트웨어가 실제로 존재하지 않았다. 확실하지는 않지만, 당시에 프레이 오토(Frei Otto - Multihalle Mannheim, 1975)와 같은 모델을 사용했던 것 같다. 반 시게루는 기계와 직접 통신하는 알고리즘을 생성한 스위스/독일 회사인 Designtoproduction과 협력할 수 있었다. 이 알고리즘을 사용하여 CNC 밀로 18m의 목재를 다듬고 약 1,800개의 이중곡선 글루렘 부분을 생산할 수 있었다. 그래서 실험적인 목재 구조가 수십년 동안 가능했음을 알 수 있지만, 알고리즘을 사용하여 실현 가능성이 커졌으며 이는 대량 맞춤화를 반영한다. 또한 대나무와 나무가 서로 다른 정체성을 가진 두 가지 재료라는 것을 인정한다. 대나무는 아시아에서 역사가 있다. 실제로 유럽이나 미국에서는 그다지 쓰이지 않지만 완벽한 건축 자재이기 때문에 그 가능성을 재고해봐야 한다. 어쩌면 유럽의 오픈 캐스트 광산 지역을 대나무로 복원하는 것이 이상적일지도 모르겠다. 빠르게 성장하는 친환경적인 대나무로 석탄을 대체하는 독특한 비즈니스 모델이 될 수도 있다.

AZC

'김나지움 (Gymnase - Sports centre in Strasbourg)'은 프랑스 스트라스부르에 있는 체육관이다. 우리는 이 프로젝트에 나무를 대량으로 썼다.

BOARD

얼마 전 우리는 보스니아 헤르체고비나의 클레코바카산(Klekovaca) 지역에 27개의 스키 슬로프가 있는 레크리에이션 스키 단지인 관광 센터를 디자인했다. 설계 과정에서 스키 슬로프를 만들려면 많은 수의 나무를 제거해야 한다는 점을 깨달았을 때, 자를 나무를 다시 사용하기로 했다. 하여 우리는 사이트의 나무에서 나온 목재를 사용하여 새로운 관광 센터를 건설할 것을 제안했다. 그렇게해서 나무를 구조뿐만 아니라 건물 파사드 대부분에도 사용할 수 있었다. 따라서 우리는 이 프로젝트를 "숲으로(Out of the Woods)"라고 불렀다. 나무의 도움과 사이트의 다양한 지리적 특징을 사용하여 3개의 다른 구역에 각 위치의 자연의 아름다움을 포용하는 3개의 다른 레이아웃과 분위기가 있는 관광 센터를 제안했다. 하나는 스키 고원과 스키 슬로프 옆에 있는 소형 메인 리조트이고, 다른 하나는 사이트 중 거의 평평한 구역에 주변의 나무처럼 자유롭게 서 있는 개별 건물로 구성된 호텔과 스포츠 단지이며, 세 번째는 소나무 숲에 있는 기후 건강 리조트 구역이다.

Carlos Lampreia

루이스 칸이 지은 목재집 피셔 하우스(Fisher House)에서 거실에 나무로 모서리 공간을 만들어 내는 방식이 놀라웠다. 외부 커버를 창문, 벤치와 다른 가구가 연결되어있는데 모두 같은 종류의 나무로 만들어진 하나의 디자인이다.

Casanova+Hernandez Architects

현재 우리 사무실은 알바니아 슈코더르(Shkodra)에 있는 호수가에 현존하는 대지에 박물관을 디자인하는 중이다. 이 박물관은 우리 사무실이 맡은 더 큰 개발지 일부이다. 사이트는 슈코더르 호수 해안 3.7km와 산 중에 위치한 해안과 평행한 새로운 보행자 및 자전거 경로로 5km가 포함된다. 새로워진 해안가와 새로운 산악길은 지역 주민과 방문객 모두에게 다양한 레크리에이션, 문화 및 관광 시설을 제공하는 10km의 보행자 및 자전거용 원형 경로를 만든다. 이 시설은 22개의 전시학적 인터벤션으로 연결된 21개의 건축 및 조경 인터벤션으로 나누어져 있다. 이러한 건축 인터벤션 중 하나는 주로 나무로 만든 이 박물관이다.

박물관의 파사드를 위해 목재를 선택함으로써 자연과 더 잘 통합할 수 있다. 파사드는 까맣게 탄 서부 붉은 삼나무 판자와 널빤지 위에 처리되지 않은 시베리아 낙엽송 막대로 만들어진다. 탄화 과정은 널빤지에 검은색의 진지한 외관을 형성하고, 그 위의 무가공 된 목재는 색상과 질감의 대비를 만든다. 이것은 시간이 지남에 따라 변할 중립적인 색상의 활기찬 외관을 만들 것이다. 이 파사드는 시간의 흐름을 반영하고 주변 환경의 색상에 더욱 더 잘 통합될 것이다.

CEBRA(Mikkel Frost)

세계에 예전 것이든 최근 것이든 멋진 목조 건물이 많이 있다. 최근에 나는 특히 쿠마 켄고가 디자인한 목조 구조물에 매료되었다. 그가 만드는 구조적 특성과 줄 패턴은 그저 놀라울 뿐이다.

나는 우리의 LIF 프로젝트에도 매우 만족한다. 이 프로젝트는 작고 약간 간과되었지만, 실제로 러시아 이르쿠츠크(Irkutsk)의 스마트 스쿨과 같이 더 큰 CEBRA 프로젝트 다수의 "어머니"이다. 여기서 구조와 클래딩은 수직 공생으로 하나가 되었고 이는 꽤 성공적이었다.

Davide Macullo Architects

우리는 예술가 다니엘 뷰렌(Daniel Buren)과 함께 스위스 산맥에 "스위스 하우스 XXXII"라는 집을 지었다. 나무 구조는 숲의 수직성을 상기시키며, 그 지역의 목재를 사용했다. 집의 원형을 한 걸음 더 발전시켰고 예술가와의 협력으로 이 건물은 조각품인 동시에 집이 되었다. 안에서 살 수 있는 조각인 것이다. 수백 년 동안 예술과 건축의 통합이 논의되어 왔고 이 건물은 두 예술이 실제로 협력하는 드문 예 중 하나이다. 이 건물은 예술이 건물의 구조적 부분이기 때문에 예술이 없으면 건물은 존재하지 않는다. 이미 다니엘 뷰런과 공동으로 설계된 벽이 있었기에 지붕의 보는 그 지역의 다른 콘셉트 예술가인 미키 탈론(Miki Tallone)과 공동으로 설계했다.

Donner Sorcinelli Architecture

높은 단열성 및 환경친화적인 디자인을 위해 CD House의 구조에 직교적층목재 패널을 사용했다. 건설 시간과 비용을 감소하고 잘 관리할 수 있었다.

Katsutoshi Sasaki+Associates

전통적인 일본 절.

Keiichi Hayashi Architect

가쇼 / 카키우치 코지

가쇼는 2011년 일본 북부에서 쓰나미에 휩쓸려 간 집의 콘크리트 기초 잔해에 만들어진 작은 피난처이다. 이는 전적으로 나무로 만들어졌으며 DIY 건설 하루 만에 설계되었다. 용감하게 행동을 취하고 나무의 특성을 창조적으로 활용한 이 건축가에게 가장 큰 존경을 표하고 싶다.

LANDÍNEZ+REY arquitectos

르 코르뷔지에(Le Corbusier)의 덜 알려진 작품 중 하나로, 프랑스 리비에라(Riviera)에 있는 로크브륀느-카프마르탱(Roquebrune-Cap Martín)에 자신의 시적인 휴가를 위해 만든 건물이 있다. 그는 나중에 이곳으로 은퇴했다.

modostudio

파르마에 있는 17세기 목조 파르네세 극장은 놀라운 걸작이다. 파르네세 일가를 위한 개인 극장으로 지어졌으며 상세한 공예의 좋은 예이다.

Mork-Ulnes Architects

스베레 펜(Sverre Fehn)의 빌라 스크레이너(Villa Schreiner)는 지지 구조와 마무리 재료로 목재를 사용하며, 사이트와 연관되고 빛을 조절하고 건물에 질감과 삶을 만드는 방식이 환상적인 프로젝트이다.

murmuro

우리는 세랄브스(Serralves) 재단을 위한 순회 아트 파빌리온에 나무를 사용했다. 구조, 벽, 창문, 문 및 가구는 모두 나무로 되어있어 간단하고 빠르게 파빌리온은 조립 및 분해할 수 있었다. 이 재료로 우리는 분해된 건물을 단일 요소로 줄일 수 있는 복잡한 연결 조인트를 설계할 수 있었다. 운송 측면에서도 빠르고 쉽고 가벼우니 전시회 사이에 필요한 창고 공간도 적었다.

NISHIZAWA ARCHITECTS

우리가 가장 좋아하는 나무를 사용한 프로젝트는 모모에다 유우 건축사무소(Yu Momoeda Architecture Office)의 농업 예배당(Agri Chapel)으로, 프랙탈 구조 시스템을 갖춘 일본 목조 예배당이다.

이 프로젝트는 일본 전통 목재 구조 시스템과 고딕 양식을 독창성있게 현대식으로 결합하기에 좋아한다. 또 다른 이유는 예배당의 활동을 원활하게 주변 자연환경과 연결하는 방식이다.

OOIIO Architecture

나는 목재를 건물의 주재료로 사용한 적이 없다. 스페인의 건조하고 매우 밝은 날씨에서 나무는 매우 흔한 건설 재료가 아니며, 외부에 사용하면 태양에 손상된다. 목재는 전통적으로 고대 건축물의 구조에서만 사용되었다.

나는 목재로 된 슬래브나 지붕 구조물을 보수하는데 이 재료를 쓴 적이 있다. 그래서 목재에 대한 나의 경험은 건물의 보수와 물리적 특성을 복구하는 목재의 처리 방법에 연관된다.

Piuarch

라테리아 소시알레 발텔리나(Latteria Sociale Valtellina)라는 오래된 건물의 재개발 프로젝트를 위해 우리는 산악 지역에서 널리 사용되는 재료인 목재를 사용하기로 결정했다. 우리는 앞면과 옆면에 돌출된 하나의 지붕을 만들었고, 파사드의 불투명한 클래딩, 구조 및 지붕의 내부에 나무를 사용했다.

SLOT STUDIO

가장 최근의 프로젝트 중 하나는 토코마데라 소매점인데, 분사 그대로 나무를 만진다는 듯이다. 바닥, 벽 및 가구에 각각 다른 마감과 니스를 사용하며 전시된 나무 상품과 어울리도록 알맞은 표현과 균형을 찾아야했던 인테리어 디자인 프로젝트였다. 따뜻함을 풍기고 상점에서 판매하는 제품을 떠올리게 수공예 느낌의 나무를 필수 요소로 설정한 프로젝트이다.

Stefano Corbo STUDIO

후지모토 소우의 목조 주택(2006년)은 건축에서 나무를 혁신적인 방식으로 사용하는 사례다. 이 프로젝트에서는 사실 바닥, 벽 및 천장 사이에는 분리가 없다. 그 형태는 350mm 정사각형 종단 삼나무 보를 불규칙하게 쌓아서 만들어졌다. 최종 결과는 구멍이 뚫린 나무 상자로 특이한 공간적 표현을 유발한다.

stpmj Architecture

Shear House /stpmj - 경량 목구조에 외부 마감까지 목재로 진행했던 프로젝트로 틀어진 박공지붕의 타이폴로지를 단일 재료인 나무로 디자인.

SUPA architects schweitzer song

우리에게 나무는 끝없이 매혹적이며 건축에서 가장 근본적인 재료이다. 우리 둘 다 나무에 둘러싸여 자랐다. 률은 한국의 한옥에서 자랐고, 나는 오스트리아 오버외스터라이히(Oberösterreich) 주의 한 농가에서 자랐다. 이러한 경험은 여전히 우리의 사고, 미학, 그리고 건축적 계획을 지배한다. 이 두 집 모두 물리적으로 멀리 떨어져 있는 것처럼 보이지만 철학적으로 우리 문화 사이에 공통된 기반을 제공했다.

TAKK Architecture

후지모토 소우의 목조 주택이다. 한 재료를 특정한 모양(나무 프리즘)으로 사용하고 특정한 방식으로 모음으로써 매우 풍부한 공간을 만든다.

TOUCH Architect

태국의 찬타부리에 위치한 이 프로젝트 이름은 찬 포 에코 스테이(Chan'4 ECO-STAY)이다. 이 건물은 사이트 자체의 특성에서 파생되었다. 찬타부리 지방에는 다양한 유형의 지형이 있으며, 특히 바다와 언덕이 주요 지형이다. 이 사이트에는 주변에 바다와 언덕이 있을 뿐만 아니라 내부에 늪과 호수, 맹그로브가 있다. 이 홈스테이에는 숙박 시설과 유기농 농장이 있는데 이 농장은 의뢰인이 은퇴 후 하는 일이다. 이 농장에서는 모든 사람이 환경과 어우러져 살며 배울 수 있고, 지내고 가는 사람들의 일부가 된다.

이 건물에는 홈스테이를 통합하는 세 층이 있다. 주 진입로는 1층에 있는 로비 공간으로 이어진다. 이 로비는 1층에 동선이 흐르는 열린 공간이 있는 태국 전통 주택의 특성인 '타이 뚱(Tai-Toon)'을 유지한다. 2층에는 각 방이 가족 여행객에게도 적합하며 호수와 바다를 볼 수 있다. 또한, 지붕에는 모든 사람이 쓸 수 있는 다목적 활동을 위한 공용 공간이 있다.

건물 전체에 사용된 재료는 나무이다. 철근 콘크리트로 만들어진 기본 구조, 기둥 및 들보를 제외한 모든 피시드 요소, 바닥, 벽, 지붕 구조에서 사용했다. 그 이유는 강철 대신에 목재를 사용하면 바다 소금 때문에 발생하는 극도의 녹을 피하는 데 도움이 되기 때문이다. 또한, 아름다움과 기능적인 면에서 봐도, 조정 할 수 있는 나무 격자 파사드로 훤히 트인 전망이 보이고 자연 환기를 할 수 있으며 건물이 자연과 조화를 이룸을 알 수 있다.

UNStudio

몇 년 전에 완성한 W.I.N.D 하우스 (W.I.N.D House)라는 한 단독주택의 파사드에 나무를 사용했다. 목재를 수열처리하여 내구성을 높였고 화학처리되지 않아 자연적으로 보인다. 시간이 지남에 따라 목재는 회색으로 바뀌며 이는 인근 해변에서 볼 수 있는 유목과 집을 둘러싸고 있는 모래 언덕을 나타낸다.

Q2
본인이 생각하는 목재의 장단점에 대해 얘기해달라.

Arenas Basabe Palacios Arquitectos

일반적으로 한 프로젝트에서 목재의 존재는 다른 건축 요소와 강한 이중성을 수반한다. 우리의 '상자 속의 상자' 프로젝트에서 봉사하는 공간을 감싸는 OSB 보드의 박력은 깨끗한 라인과 뉴트럴한 톤의 재료(회색 광택 콘크리트 바닥과 반투명, 투명, 그리고 불투명한 벽으로 된 흰색 팔레트)로 된 실내 경기장, 강당, 예배당, 교실, 회의실과 사무실 같은 주요 공간과 크게 대조된다. 이 이중성은 내부의 나무 상자와 기존 파사드를 감싸는 새로운 금속 껍질 사이에서도 볼 수 있어 '상자 안의 상자'에 대한 아이디어에 더 의미가 생긴다.

ARPHENOTYPE

나무는 자연이 매일 대량으로 보충한다. 지속가능한 임업 및 책임있는 소비로 모든 건설 목적을 위한 목재는 항상 충분할 수 있다. 목재는 적절하게 가공하고 의도적으로 사용한다면 다른 재료 대부분의 내구성을 훨씬 뛰어 넘는다. 지난 200년에서 500년 사이에 지어진 목조 건물 일부는 종종 여전히 서 있다. 그러나 나무는 가연성이 높으며 해충, 박테리아 또는 곤충에 의해 공격받을 수 있다. 대나무는 무엇보다도 그 엄청난 성장 속도가 아주 인상적이다. 대나무는 험한 상황에서도 내구성이 강하고 튼튼하다. 또한 세포 구조의 구멍 덕분에 매우 가볍고 탄력적이다.

AZC

나무는 이산화탄소가 낮은 환경 친화적인 재료이고, 매우 따뜻하며 인간적이다.

하지만 외관에 사용할 때는 보호 처리가 필요하고 주요 구조로 사용할 때는 화재 방지 처리가 필요하다.

BOARD

수천 년 동안 인간이 가장 좋아하는 건축 자재 중 하나로 꼽힐 만큼 나무에는 많은 장점이 있다. 그 장점 중 하나는 의심할 여지없이 풍부한 가용성, 즉 공급량을 갱신하는 능력이다. 또한 특히 세로 방향의 모양, 강인한 동시에 가벼운 점, 모양을 바꾸기 쉬운 점, 그리고 대부분 나무가 부드러워 이음매를 쉽게 만들 수 있는 점 덕분에 나무는 집, 보트, 가구 등 모든 것을 만드는 데 좋은 재료가 된다. 하지만 지적하고 싶은 나무의 단점 중 하나는 벽돌, 타일 혹은 유리처럼 특정한 장소에서 생산할 수 없다는 것이다. 목재를 사용하기 전에 나무가 오랫동안 자랄 필요가 있고 수확과 운송에는 많은 어려움이 있다. '숲으로' 프로젝트에서 우리는 새로운 관광센터 건설에 바로 그 자리에서 나는 목재를 사용함으로써 목재 운송과 관련된 어려움을 피하고 싶었다.

Carlos Lampreia

나무와 흙은 인류가 피난처를 짓기 위해 사용한 최초의 재료이다. 나무로 우리는 집에서 가구까지 거의 모든 것을 만들 수 있다. 이런 의미에서 집에 나무가 있는 것은 마치 친구를 집에 둔 것 같다. 목재는 매우 유기적이고 살아있는 재료이며, 움직이고, 소리와 향기가 있기 때문에 인공 나무로 생각할 수도 있다. 유일한 문제는 숲이 필요하기 때문에 도시 전체를 나무로 만드는 것이 불가능하다는 점이다. 이 부정할 수 없는 현실은 나무를 조심스럽게 사용해야 하는 비싸고 희귀한 재료로 만든다.

Casanova+Hernandez Architects

FSC 마크가 있는 목재를 사용하는 것은 친환경적이며 건물의 탄소 발자국을 최소화하는 데 도움이 된다. 자연의 법칙을 따라 시간이 지나면서 나무의 색과 질감은 변할 것이며, 방문객에게 이 지역에 서식하는 새의 둥지를 상기시키고 나무의 향은 호수와 인접한 산에 과거에 존재했던 숲의 시적 공명을 깨울 것이다.

한편으로 나무는 편안함을 제공하고 시적 의미가 넘치며, 자연을 상기시키고, 시각적 특성, 촉감 및 깊은 냄새 덕분에 치유 경험을 자극할 수 있는 재료이다.

다른 편으로 친환경적인 재료인 목재는 디자이너가 자연에 대한 존중하에 보다 지속 가능한 환경을 조성하는 데 도움이 된다.

CEBRA(Mikkel Frost)

나무로 많은 층을 지으려면 수년 동안 소방법이 난제였다. 많은 의뢰인이 목재에 지속적인 관리가 필요하다며 유지 보수에 대해 불평하지만, 나는 그에 동의하지 않는다. 여러 면에서 나무는 환상적인 재료이다. 거의 모든 것에 사용할 수 있다. 나무는 저렴하고, 작업하기 쉽고, 구조와 공전적인 면에서 매우 좋다. 재미있는 것은 나무는 숲에서 그냥 자라는 것뿐이라는 점이다. 그냥 잘라내면 사용할 준비가 거의 되어있다. 구울 필요도, 섞을 필요도, 강화할 필요도 없다. 더 필요하면 그냥 심고 기다리기만 하면 된다. 자라는 동안 나무를 보는 것은 즐거운 일이고 새를 위한 집이 된다. 심지어 지구를 위해 산소까지 생산한다!

위에서 언급했듯이 모든 상황에 이상적인 재료는 없고 나무 역시 그렇지 않지만 완벽함에 꽤 가깝다. 나는 얼마 전 내 마당 전체를 원목 마루로 깔았다. 마루는 맨발로 걷기 아주 좋고 햇볕에 적당히 따뜻해진다.

Davide Macullo Architects

나무는 그 용도에 있어서 자연과 인간에 가장 가깝기 때문에 가장 아름다운 재료 중 하나이다.

Donner Sorcinelli Architecture

나무는 우리가 재료에 요구하는 모든 자질의 합이다. 하중을 운반할 만큼 강하며, 시각적 및 촉각적인 면에서 자연스럽고, 재활용할 수 있으며 따뜻하다.

Katsutoshi Sasaki+Associates

장점: 부드럽다는 점이다.

단점: 단점도 부드럽다는 점이다.

Keiichi Hayashi Architect

장점: 모든 면에서 다루기 쉽고 감촉이 좋다.

단점: 쉽게 탄다.

LANDÍNEZ+REY arquitectos

우린 우리 주변 환경의 재료를 사용해야 한다. 지구가 우리에게 주는 우리 주변에 있는 재료말이다. 우리는 건축에서 나무를 사용하는데 어떤 단점도 못 찾았다. 목재를 사용하는데 단점이 있다면 순전히 문화적인 이유거나 재료의 부재와 관련이 있다.

modostudio

나무는 다양한 상황에서 사용할 수 있는 재료이다. 구조용 건설 자재가 될 수도 있고, 장식용 재료가 될 수도 있다.

나무는 매우 유연하며 자연과 강한 연관성이 있다.

Mork-Ulnes Architects

나무는 경제적이고 지속 가능하며 인간의 감각(촉각, 후각, 시각)에 독특한 호소력을 지녔고 아주 다용도이다. 유일한 실질적인 제한은 매우 큰 구조이거나, 물과 직접 접촉하는 경우, 혹은 특정 방화 조건 하이다.

murmuro

나무는 상상할 수 있는 거의 모든 디테일을 만들 수 있는 아주 굉장한 재료이다. 따뜻하고 편안한 느낌을 주고 사람들과 마음이 통한다. 나무의 가장 큰 약점은 노후화를 줄이기 위해 지속된 관리가 필요한 점이다. 나무 건물은 디자인과 관리가 잘 된다면 수세기 동안 오래 갈 수 있다.

NISHIZAWA ARCHITECTS

나무의 장점은 콘크리트 없이 건물 전체의 구조로 사용할 수 있다는 점이다. 그 외에도 천연 목재의 색상과 질감은 건물 자체의 감정을 보여주는데 도움이 된다. 우리에게 나무의 단점은 여러 환경에서 사용하기 어려운 민감한 물질이기 때문에 온도, 습기, 곤충 감염으로 쉽게 손상된다는 점이다. 두 번째는 구조적 하중이 사용하는 목재의 종류에 달려 있기 때문에 때로는 지지력을 증가시키기 위해 구조의 일부를 강철과 결합해야 한다는 점이다.

OOIIO Architecture

장점: 훌륭한 구조적 속성이 있고 재활용 가능한 재료이며 지속 가능한 숲 (FSC 숲)의 목재로 작업하면 그 생산 역시 지속 가능하다. 생산(자라는) 과정에서 산소를 생산하는 유일한 재료이다.

단점: 기생충의 영향을 받을 수 있으며, 불에 약하고, 내가 일하는 건조하고 더운 날씨에 쓰기에 좋은 재료는 아니다.

SLOT STUDIO

나무는 고귀함, 따뜻함, 장인 정신을 지닌 매우 특별한 재료이다. 나무에는 특별한 개성이 있고 그 매력을 드러내는데 어떤 과정도 필요하지 않기 때문에 우리는 프로젝트에 나무를 사용하는 것을 좋아한다. 나무는 치수, 가단성, 질감 및 음조를 달리 하는 것만으로도 매우 다양한 느낌을 준다. 또한 지속 가능하고 융합적인 소재이다.

Stefano Corbo STUDIO

나무에 대한 장단점을 정의하는 것은 어렵다.

모든 재료는 프로젝트의 전반적인 논리는 물론 다른 요소 간의 관계에 따라 고려되어야 하기 때문이다.

stpmj Architecture

장점: 가볍고 시공이 용이하며 자연친화적이다. 어떤 처리를 하느냐에 따라 시각적인 효과와 더불어 촉각적, 후각적 감각들을 동시에 느끼게 할 수 있는 천연재료

단점: 기후에 따라 수축 팽창에 따른 변위에 대한 고려가 설계 및 시공 시 요구된다.

SUPA architects schweitzer song

건축 자재로서 나무의 압도적인 장점은 형이상학적 특성이다. 나무 안에서 살 때 나무는 우리 자신의 삶을 반영하며 같이 살아간다. 나이를 먹고, 움직이고, 소리를 내고, 따뜻하고, 냄새가 나고, 변화한다.

TAKK Architecture

목재는 우리가 구조와 외관 모두에 많이 사용하는 재료이다. 값이 싸고 친환경적이며 열적 특성이 매우 좋다.

TOUCH Architect

모든 물질에는 나름의 약점이 있다는 것을 부인할 수 없으며 나무도 마찬가지이다. 경재(hard wood) 만이 구조에 사용될 수 있지만 비싸고 찾기 어려우며, 연재(soft wood)는 내부 마무리, 가구 및 클래딩에만 사용할 수 있으며 흰개미에 저항력이 없다. 연재 중 일부는 늘어나거나 수축하기 쉬워서 문이나 창틀로 사용하기에 적합하지 않다.

단점 외에도, 수많은 나무의 장점은 이미 위에 언급했다. 물리적으로 강하여 건물의 기본 구조에 적합하고 콘크리트 및 강철과 같은 다른 구조와 비교해 가볍고 유연하다. 지속할 수 있고 환경친화적인 건축 자재이며 유일하게 재생할 수 있다. 또한 나무는 구조에만 사용되지 않고 자연과 조화로운 느낌을 전달하는 마무리에도 사용된다. 부드러워 보이고 스타일 있게 편해 보이기도 한다.

UNStudio

사람들은 종종 나무가 충분히 오래 지속되지 않거나 유지를 위해 너무 많은 보수가 필요하다고 걱정하지만 실제로 오늘날에는 나무의 내구성을 매우 높일 수 있는 처리법이 있다.

Interviewee PROFILE

Arenas Basabe Palacios Arquitectos

Overview

Arenas Basabe Palacios arquitectos is a young architecture and urbanism studio based in Madrid. Its partners Enrique Arenas Laorga, Luis Basabe Montalvo and Luis Palacios Labrador have been working together since 2006 and have won more than thirty prizes in architecture and urbanism competitions. They have given lectures in diverse institutions and presented their work and investigation in several international exhibitions. Their work has been published in Spain, France, Italy, Switzerland, UK, Austria, Germany, Cyprus, India and Korea.

Directors

Enrique Arenas Laorga (1974), architect (ETSAM, Madrid) and Doctor Cum Laude (UPM Madrid, 2016). He has developed projects in very different areas: rehabilitations, housing, institutional and events. He has held lectures at several academic institutions, and taught as professor at the European Institute of Design in Madrid (IED).

Luis Basabe Montalvo (1975), architect graduated at the TU Graz. He has been visiting professor at Dipartimento di Architettura e Studi Urbani (DAStU) at Politecnico di Milano. Since 2003 he teaches design studio at ETSAM as an associate professor. He has been guest researcher and guest lecturer at various Universities: RWTH Aachen (Germany), Cambridge (UK) and CEPT Ahmedabad (India).

Luis Palacios Labrador (1983), architect graduated at ETSAM (Madrid, 2009), Master in Advanced Innovation and Technology (ETSAM, 2011) and Doctor Cum Laude (UPM, 2017). He currently teaches design studio as an associate professor. He has worked in the Netherlands, investigated in Berlin and held lectures and workshops in India and UK.

ARPHENOTYPE

Dietmar Köring, Dipl.-Ing.(FH) M.Arch. Architect BDA, is an architect, researcher, and educator living in Cologne. He is head of the architectural research office Arphenotype, where he focuses on blurring the boundaries of different artistic disciplines. From 2012 to 2017 he was a research fellow at TU Berlin / CHORA City & Energy and Dietmar has taught Digital Design at TU Braunschweig from 2010 to 2012, he was Guest Professor for Virtual Realities & Experimental Architecture at the University Innsbruck ./Studio3 in 2011, Technology and Design Lecturer at the Cologne Institute for Architectural Design / C-I-A-D and visiting lecturer for digital design at the DeMontfort University Leicester. From 2011 to 2012 he was assistant professor for Smart Grid research (Smart City Concepts 2022) at the Institute for Corporate Architecture at the Cologne Technical University.

He studied architecture at the University of Applied Sciences Cologne, the University of Western Sydney and at the Muthesius Academy of Fine Arts, where he graduated as in 2005 as Dipl.-Ing. (FH). Dietmar received his MArch in 2007 at the Bartlett School of Architecture University College London, under Prof. Neil Spiller and Phil Watson. Since 2008 he is a registered Architect at the AKNW and ARB.

Through his career he has worked internationally for offices such as Coop Himmelblau, Graft, 3deluxe and Andrew Wright Associates. His research has been awarded by the Jaap Bakema Fellowship / NAI and his works have been internationally published and exhibited, including MoMa New York, Heide Museum of Contemporary Arts Australia and Deutsches Technikmuseum Berlin. Dietmar has given international lectures, guest critiques and workshops. Since 2013 he is collaborating with Simon Takaski as Takasaki Koering Architects.

Dietmar is member of the narrative research network .horizon.com.

AZC

AZC was founded in 2001 with the idea that exploring architecture and its techniques could help to improve our built environments. Our interest does not lie in inventing concepts, we have always sought to realize buildings for real life's needs.

Through competitions and direct commissions, our office has worked on over a hundred projects of varied scales and uses. Most of our built projects are intended for a wide audience; sports facilities, lecture halls, office buildings and residential, some of which very specific for vulnerable populations. We also have, eight metro stations under construction, including four in Paris and four in Rennes and studies for a new station in Lyon, are ongoing.

Through some recently completed buildings, which have different purposes, we want to share our current concerns of coherence with global and local contexts which today represent the major issues of architecture.

We are not alone in the projects process, our clients and our partners share this common experience, which is engaging and meaningful; they allow us to reflect on our own actions that relate to the projects. We aspire to a high quality in any form of collaboration.

Most of our work has been published, displayed, sometimes awarded and we have often been given the opportunity to speak on topics of sustainability, diversity and innovative techniques, which all illustrate our commitments.

BOARD

BOARD (Bureau of Architecture, Research, and Design) was founded in Rotterdam in 2005 and is active in many fi elds: as an architecture, urban design, and design practice, as a research board and as a platform for comparative analysis on urban issues through its bi-annual journal MONU - Magazine on Urbanism. BOARD won several prizes recently in prestigious international architecture and urban design competitions.

Bernd Upmeyer is the founder of BOARD and editor in chief of MONU - Magazine on Urbanism. He studied architecture and urban design at the University of Kassel(Germany) and the Technical University of Delft (Netherlands). From 2004 until 2007 he taught and did research as Assistant Professor at the department of Architecture, Urban Planning and Landscape Planning at the University of Kassel. In 2010 he taught as Adjunct Professor at the department of Urban Design at the Hafen-

City University Hamburg. In 2012 he was a guest critic at the Berlage Institute's fi rstyear postgraduate research studio "Anarcity".

In 2013 he lectured and participated in a discussion about architecture, urbanism and media at Strelka's Urban Studies Session in Moscow. Upmeyer frequently writes for international publications and magazines. He holds a PhD (Dr.-Ing.) in Urban Studies from the University of Kassel(Germany). Upmeyer is the author of the book Binational Urbanism - On the Road to Paradise. The book examines the way of life of people who start a second life in a second city in a second nation-state, without saying goodbye to their fi rst city.

Upmeyer coined the term "binational urbanism".

BOARD employs an international team of architects and planners and collaborates with national and international external consultants and specialists.

Carlos Lampreia

Carlos Lampreia, architect (1990), is architecture design teacher at FAA-Universidade Lusíada de Lisboa since 1994, studied at OPorto Architecture School and at Lisbon Technical University FA-UTL. Master in architecture theory, 'towards an objective architecture', 2002. Phd about, strategy, site and material, concerning architecture and arts, 'concept site and material, a strategy in architecture and arts, 1960-2000', 2017. His Lisbon based office, carloslampreia[x]arquitectos, works on an experimental way with young architects and students towards architectural materialisation, participating both in international competitions and individual private requests.

Casanova +Hernandez Architects

Casanova+Hernandez, founded in 2001 by Helena Casanova and Jesus Hernandez, is a design and research studio based in Rotterdam. It focuses on rethinking and designing our urban habitat in order to create vibrant cities while promoting environmental and social sustainability.

Working with an interdisciplinary team and with experience developing projects in very different cultural contexts in Europe, South America and Asia, the office has expanded its capabilities and its international network through close and fruitful collaboration with experts in different continents.

Casanova+Hernandez is structured in two complementary platforms: C+H Projects and C+H Think Tank. C+H Projects is the design platform of Casanova+Hernandez. It operates in the fields of architecture, landscape architecture and urban design, often combining them to create hybrid architectural landscapes.

C+H Think Tank works as an independent platform that analyses urban and social problems and proposes innovative design solutions, new urban strategies and advice on the implementation of new policies.

www.casanova-hernandez.com

CEBRA

CEBRA is a Danish architectural office founded in 2001 by the architects Mikkel Frost, Carsten Primdahl and Kolja Nielsen. In April 2017, architect MAA Mikkel Hallundbæk Schlesinger entered the group of partners.

Based in Aarhus in Denmark and in Abu Dhabi in the UAE, CEBRA employs a multidisciplinary international staff of 50 architects, constructing architects, urban planners and landscape architects, who all share a strong passion for architecture.

CEBRA has gained recognition through award-winning projects such as The Iceberg at the habour front in Aarhus and the Experimentarium science centre in Copenhagen and has a growing international portfolio in Europa and the MENA region.

At CEBRA we want to change the way to think, design and build architecture. We are always pushing artistic and architectural boundaries - pushing these boundaries with a CEBRA attitude and a Nordic mindset that combines our artistic approach to architecture with an understanding of its cultural context.

We design architecture by listening to and understanding our users and clients and studying their context, culture and climate. Our services cover all project phases - from client advisory and user involvement and concept and project development to project and construction management as well as technical supervision.

Most CEBRA projects are within the fields of education, culture and housing - thought, designed, and built in line with our mantra - Architecture with attitude.

Davide Macullo Architects

Davide Macullo (b. Giornico, CH, 1965) lives and works in Lugano, Switzerland. Studied art, architecture and interior design. For 20 years (1990-2010) he was project architect in the atelier of Mario Botta with responsibility for over 200 international projects worldwide. He opened his own atelier in 2000.

The ethos of the studio is one of 'drawing from context' and the various contributions promote a dialogue between the specificity of the project and the universality of the contexts. His work has been published and awarded both at home and abroad. Selected realized projects include the WAP ART foundation mixed use gallery and apartment in Gangnam Seoul, South Korea, the Assuta Hospital in Ashdod, Israel, 5* Hotel and SPA facilities in Greece,the headquarter Jansen AG in Oberriet, Switzerland, Private Museum in Jeju South Korea, Sino-Swiss centre in Tianjing China, several houses and housing in Switzerland and abroad.

Current projects include a new Health and Wellness Hotel in Weggis, Switzerland and Marbella, Spain, houses and residential buildings in Switzerland, a beachfront villa in Heraklion, Greece, a Medical SPAin Baku, Azerbaijan. The work of the studio includes masterplanning, graphic design, branding consulting and custom designed furniture, now in production and spans to the creation of contemporary art collections for clients.

In Rossa Calanca Valley in the Grison Canton, Davide Macullo has started an urbanistic program to promote the intervention in situ of international artists to influence daily life through contemporary art. The first building realized in collaboration with Daniel Buren will be followed by other ten artists.

Donner Sorcinelli Architecture

Donner Sorcinelli Architecture is an international architectural design office based in Italy.

Founded by architects Luca Donner and Francesca Sorcinelli, the firm pays particular attention to the theme of sustainable and affordable architecture in all its variants, based on experimentation and research in various fields like Architecture, Urban Design, Interior and Product Design.

Their projects have been awarded in International competitions:

"Social Housing Dev."- Piazzola sul Brenta 1st prize; " Social Housing Dev." -Presina, 1st prize; "Design Beyond East and West"- Seoul, 1st prize; "International Design Competition for Modern Saudi Houses, Affordability and Sustainability"- Riyadh, 1st prize; Urban Retrofitting of S.Elena's - Silea, 1st prize; Sansovino Masterplan -Montebelluna, 3rd prize; School Campus in Carbonera, 3rd prize;"Your Absolute" for a residential Tower, Mississauga, Honorary Mention; "Daejeon Urban Renaissance"- Daejeon, Honorable Mention.

They have been published in many international magazines, books and presented in several exhibitions in Italy and abroad.

DoSo are winners of the "Cityscape Architectural Review Award 2006", "SAIE selection Awards 2009" and the "20+ 10+ X World Architecture Award 2012". They have received an Honorary Mention at Modern Atlanta Prize 2011, an Acknowledgement Prize at Holcim Awards 2005 for sustainable constructions (MENA region) and they have been nominated by Korean Institute of Architects among "100 Architects of year 2017".

Luca Donner and Francesca Sorcinelli have been teaching at International Universities in Dubai after previous academic experiences in Italian Universities.

www.doso.it

Katsutoshi Sasaki +Associates

Office Information
4-61-3 Tanaka-cho
Toyota-shi Aichi
471-0845 Japan
Phone:+81 565 29 1521
sasaki@sasaki-as.com

1976 Born in Toyota-shi Aichi,Japan
1999 Guraduated Kindai University
2008 Established Katsutoshi Sasaki + Associates

Keiichi Hayashi Architect

Office information
1-6-10 Kumochi-cho Chuou-ku
Kobe-city Hyogo
651-0056 JAPAN
Phone +81.78.221.1868
hayashi @8107.net
Owner: Keiichi Hayashi
Established: 1997

Biography
1967 Born in Osaka
1991 Graduated from Metal Engineering, Kansai University
1993 Graduated from Architecture, Kansai University
1997 Established Keiichi Hayashi Architect

Design Philosophy
It is important for me to make architecture using basic materials and uncomplicated construction methods. I try to create a system that is based on pure architecture but becomes complex when people use it.

www.haya-at.com

LANDINEZ+REY arquitectos

LANDINEZ + REY is an architectural practice co-founded in 2000 and based in Madrid(Spain) by the architects David Landinez González-Valcácel (Madrid, 1973) and Mónica González Rey (Paris, 1973). Both are formed as M.Arch (1999) in ETSAM-UPM(Faculty of Architecture of the Polytechnical University of Madrid, UPM) and both are also graduated as Building Engineers by UEM-Madrid (2013). David also is M.Arch in Efficient Buildings and Rehabilitation by UEM (Universidad Europa de Madrid) and Mónica has also postgraduated studies in Analysis and Real Estate Management by Colmillas University (ICAI-ICADE)

LANDINEZ + REY arquitectos [el2gaa] develops its activity as a working plaform where architecture is sought from its capacity as a system generator. Systems capable of providing answer to both the place and the rest of the dimensions, scales and techniques demanded by each draft. Its work is based on the research that arises from the proposals presented to architectural competitions, many of them has been rewarded with awards and honorable mentions, others won and built.

Among these works we can find the Badajoz, Coria and Plasencia secondary schools and the Malpartida de Plasencia and Jaraíz de la Vera Gym Plvillions for the Regional Government of Extremdura, Spain and the RivasFutura Underground Station for Metro de Madrid.

His works have been published in different specialized magazines and exhibitied by different national and foreign insititutions.

www.landinez-rey.com

modostudio

modostudio | cibinel laurenti martocchia architetti associati, located in Rome, is a multidisciplinary practice of architecture, urban planning and industrial design.

Profiting from the diversified skills of the founding partners and the continual collaboration with experts from various fields, modostudio combines architectural theory, research, innovation and experimentation with high technical knowledge and professionalism. Established at the end of 2006 by three principal architects, Fabio Cibinel, Roberto Laurenti, and Giorgio Martocchia, after many years of collaborating with internationally acclaimed architects like Massimiliano Fuksas, Piero Sartogo, Erik Van Egeraat and Kas Oosterhuis, modostudio in a short time was awarded and shortlisted in many international architectural competitions. The office completed a 13.500sqm office building in Nola (IT) for Giorgia & Johns spa fashion company, the new headquarters for the Elisabeth and Helmut Uhl research foundation in Laives (IT) and a research building for Intecs spa company in Rome. Actually is ongoing the desing for new theater and urban park of Piazza d'Armi site in L'Aquila (IT).

Mork-Ulnes Architects

About Casper Mork-Ulnes

Norwegian born, Casper Mork-Ulnes was raised in Italy, Scotland and the United States, which has brought a broad perspective to his eponymous firm's work. In 2015, Casper was named one of "California's finest emerging talent" by the American Institute of Architects California Council. He was selected by the Norwegian National Museum as one of "the most noteworthy young architects in Norway" with the exhibit "Under 40. Young Norwegian Architecture 2013." Casper holds a Master of Architecture from Columbia University and a Bachelor of Architecture from California College of the Arts.

With offices in San Francisco and Oslo, Mork-Ulnes Architects approaches projects with both Scandinavian practicality and Northern California's 'can-do' spirit of innovation.

Rigorous and concept-driven, the practice is based on built work characterized by both playfulness and restraint, and informed by economies of means and materials. Mork-Ulnes Architects have worked on projects ranging in scale from masterplans to 100 square foot cabins, and have realized buildings on 3 continents.

Mork-Ulnes Architects has been the recipient of numerous national and international honors, including Architectural Record's 2015 worldwide Design Vanguard award.

The work of Mork-Ulnes Architects has also been widely featured in international publications such as The New York Times, Wallpaper, Mark, and Dwell.

www.morkulnes.com

murmuro

João Caldas (Braga, 1981) is the co-founder, with Rita Breda (Estarreja, 1981) of the office murmuro, working from Braga and Porto, in Portugal. They have started their joint path while still at the university, having graduated from DARQ-FCT, University of Coimbra. As Erasmus students, at the NTNU and the Fine Arts Academy in Trondheim, Norway, they had the opportunity to taken part of an interdisciplinary and collaborative program for architecture and art students, pivotal in their education.

The success of some of their projects led to the foundation, in 2015, of the office murmuro, where they develop projects regardless of their scale, program or budget. From their body of work they highlight the multifamily housing project Doze Casas, in Braga, shortlisted for the PREMIS FAD award in 2016, the VII ENOR Award in 2017, and the 2nd Prize on the Serralves Foundation's Pavilion Competition. Recently murmuro has received EUROPE 40 UNDER 40 award that spotlights the most promissing and the best emerging young architects in europe.

NISHIZAWA ARCHITECTS

SHUNRI NISHIZAWA
was born in 1980. He obtained his B.Arch (2003) and M.Arch (2005) degrees from Tokyo University. He worked from 2005 to 2009 for Tadao Ando Architect & Associates (Osaka). Then, he worked with Vo Trong Nghia (Ho Chi Minh City, 2009-11). He was a Partner in Sanuki+Nishizawa Architects (2011-15), before founding Nishizawa Architects in 2015. His work includes the Binh Thanh House (with Vo Trong Nghia Architects, Ho Chi Minh City, 2013); Thong House (Ho Chi Minh City, 2014); Katzden Factory (Binh Duong City, 2016); House in Chau Doc (2017; published here); Pizza 4P's Ben Thanh (Ho Chi Minh City,2017), all the completed buildings are in Vietnam.

OOIIO Architecture

OOIIO is an international team of architects, designers and engineers engaged in finding this special "I don´t know what it is" that makes a work unique, exciting and able for transmit sensations on a way that a vulgar work will never get.

In OOIIO we know that not every construction is architecture.
We do architecture.

OOIIO creative process is completely open and random. When we start a project we never know how it will end, what will come out, that's why we introduced throughout the design process countless stimulus and references from any kind that may improve the final design. These are usually unexpected and we find them everywhere, like in the geometry of a mineral that seems attractive, a traditional dish of the place where the project is built, the shape and colors of a vase, a tree or a graffiti that we found painted on a wall.

This constant search for poetical links to architecture through everything around us makes us look at the world with wide open eyes and helps us to get every project as the result of a unique creative process, making each OOIIO project special and standed out.

www.ooiio.com

Piuarch

Francesco Fresa, Germán Fuenmayor, Gino Garbellini and Monica Tricario formed the Piuarch studio in 1996 out of a desire to merge different experiences into a shared architectural project.

The studio is located in an open space in a former industrial building that once hosted a typography business in Brera, in the centre of Milan. Here, Piuarch designs public buildings, office and residential complexes, commercial spaces, boutiques, shopping malls and even urban plans, with the contribution of consultants from various disciplines.

Piuarch has pursued these themes participating in competitions, developing projects from the planning to the final construction phase, elaborating interior design projects.

In recent years Piuarch has developed a number of projects abroad. It is active in China, Algeria, Russia, where it has recently opened an operational office, and in Ukraine, with ongoing and realized projects.

SLOT STUDIO

SLOT is an active and interdisciplinary architectural design studio. People from diverse professional disciplines contribute to this project.

Our work has reached a profound understanding of the human needs, enabling us to intertwine the constructive and philosophical sides of building.

As architects, we push design to its ultimate material consequences and aim for cultural connectivity: sense of usability, mathematics of space and a wide aesthetic research.

Juan Carlos Vidals, Founder and Director of SLOT, graduated from the UNAM - National Autonomous University of Mexico in 2002 and has worked in the LCM/Fernando Romero and collaborated with Rojkind Arquitectos in several projects.

slot.mx

Stefano Corbo Studio

Stefano Corbo (1981) is an Italian architect, researcher, and Assistant Professor at RISD (Rhode Island School of Design).

He holds a Ph.D and an M.Arch. II in Advanced Architectural Design from UPM-ETSAM Madrid (Escuela Técnica Superior de Arquitectura).

Stefano has taught at several academic Institutions: Nanjing University; LAU Beirut (Lebanese American University); The Faculty of Architecture in Alghero, Italy; ETSAM Madrid; he has also been a guest lecturer at SAC Städelschule Frankfurt, Deakin University in Melbourne, College of Design Minnesota, ESALA Edinburgh, The University of Miami, and The University of Wisconsin.

Stefano has contributed to several international journals and has published two books: "From Formalism to Weak Form. The Architecture and Philosophy of Peter Eisenman." (Ashgate-Routledge, 2014), and "Interior Landscapes. A visual atlas." (Images, 2016).

In 2012, after working at Mecanoo Architecten, Stefano founded his own office SCSTUDIO (www.scstudio.eu), a multidisciplinary network practicing architecture and design, preoccupied with intellectual, economic and cultural contexts.

www.scstudio.eu

stpmj

stpmj is an award winning design practice based in New York and Seoul. The office is founded by Seung Teak Lee and Mi Jung Lim with the agenda, "Provocative Realism". It is a series of synergetic explorations that occur on the boundary between the ideal and the real. It is based on simplicity of form and detail, clarity of structure, excellence in environmental function, use of new materials, and rational management of budget. To these we add ideas generated from curiosity in everyday life as we pursue a methodology for dramatically exploiting the limitations of reality.

We design identity, brand and value.
We design creative and innovative culture.
We design unique vision of architecture.

We have been recognized with architectural awards including,

Architectural Record Design Vanguard

American Institute of Architects New York Design Award

Korean Ministry of Cultre, Sports and Tourism Young Architects Award

Kim Swoo Geun Preview Award

American Institute of Architects New Practices New York

Architectural League Young Architects + Designers

SUPA architects schweitzer song

Ryul Song was born in Taejon, Korea, and graduated from Hong-Ik University Seoul and Technical University of Dortmund, Germany. Christian Schweitzer was born in Linz, Austria, and graduated from Kaiserslautern University of Technology, Germany. Both met during their master's degree studies at the Frankfurt Staedelschule Academy of Fine Arts and established in 2000 their studio SUPA architects schweitzer song in Frankfurt, Germany. In 2005 they branched out to Seoul, Korea, teaching at Korea National University of Arts, Seoul National University, Korea University, and Hanyang University.

SUPA is working within the narrow intersection of conceptual design, art, theory and education. They are focused on the conceptual approach towards architectural design, finding new ways to expand the vocabulary of architecture. Their special interest lies in the exploration of the specific sociocultural context inherent in a task as the interface between everyday life, art and architecture.

TAKK Architecture

Takk is a space for architectural production focused in the development of experimental and speculative material practices in the intersection between nature and culture in the contemporary framework, with a special attention on the overcoming of anthropocentrism on its different ways (political, ecological, cultural, on gender), and on the definition of new notions of beauty through the articulation of the difference by assembling a multiplicity of materials from different origins and conditions.

Additionally to this profesional practice, Takk is developing a framework in the field of research and teaching. Mireia Luzárraga and Alejandro Muiño are teachers in the Projects Department of the Universidad de Alicante (UA), in BAU Barcelona Design University Centre, and Master Tutors in the Institute of Advanced Architecture of Catalonia (IAAC). They have also participated as teachers in different workshops and summer schools and have explained their work in several lectures internationally.

At the present time, Mireia and Alejandro combine their profesional and teaching labour with the development of their respective PhD Thesis on the politics of ornament and self sufficient micro-communities. They have been granted for them with the scholarship "Junior Faculty - La Caixa".

TOUCH Architect

TOUCH Architect Co.,Ltd. was first established since 2014. It was changed from TOUCH STUDIO Architect partnership, with four years experiences into a company. With a great chance of an improvement, we have two main co-founders consist of Mr. Setthakarn Yangderm as an architect and leader of our firm and Ms. Parpis Leelaniramol as an architect, which will corporate together in design, construction, and management.

SETTHAKARN YANGDERM
Architect // Managing Director // Founder
Bachelor of Architecture
(Department of Architecture, Major in Thai Architecture)
Faculty of Architecture, Chulalongkorn University. (1st top rank of Thailand's university)
Honor in best architectural design in 2007.

PARPIS LEELANIRAMOL
Architect // Co-founder
Bachelor of Science
(International Program in Design and Architecture),
Faculty of Architecture, Chulalongkorn University. (1st top rank of Thailand's university)
Master of Science in Real Estate Business (MRE),
Faculty of Commerce and Accountancy, Thammasat University. (1st top rank of business program)
Teaching Assistant - MRE Thammasat Personal Consultant - Mini-MRE at Ananda Development (Real Estate Listed Company in Thailand)

UNStudio

Ben van Berkel studied architecture at the Rietveld Academy in Amsterdam and at the Architectural Association in London, receiving the AA Diploma with Honours in 1987.

In 1988 he and Caroline Bos set up an architectural practice in Amsterdam, extending their theoretical and writing projects to the practice of architecture. UNStudio presents itself as a network of specialists in architecture, urban development and infrastructure. Current projects include the design for Doha's Integrated Metro Network in Qatar, 'Four' a large-scale mixed-use project in Frankfurt and the Wasl Tower in Dubai.

With UNStudio he realised amongst others the Mercedes-Benz Museum in Stuttgart, Arnhem central Station in the Netherlands, the Raffles City mixed-use development in Hangzhou, the Canaletto Tower in London, a private villa up-state New York and the Singapore University of Technology and Design.

In 2018 Ben van Berkel founded UNSense, an Arch Tech company that designs and integrates human-centric tech solutions for the built environment.

Ben van Berkel has lectured and taught at many architectural schools around the world. Currently he holds the Kenzo Tange Visiting Professor's Chair at Harvard University Graduate School of Design, where he has led a studio on health and architecture. In 2017, Ben van Berkel also gave a TEDx presentation about health and architecture. In addition, he is a member of the Taskforce Team / Advisory Board Construction Industry for the Dutch Ministry of Economic Affairs.

Architect

se study of

W.I.N.D House, Noord-Holland, The Netherlands
UNStudio

landscape becoming building

switch

Concept

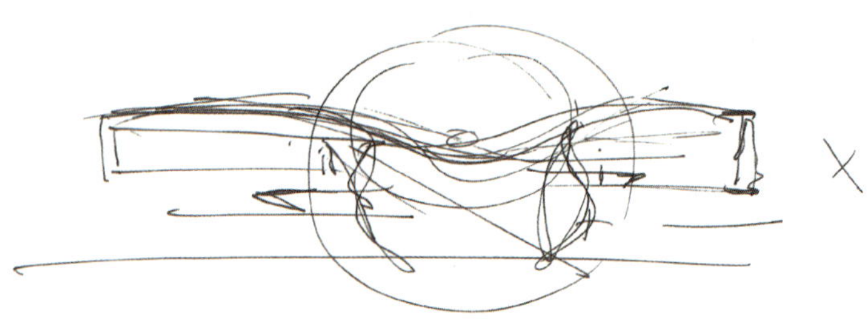

Concept Sketch

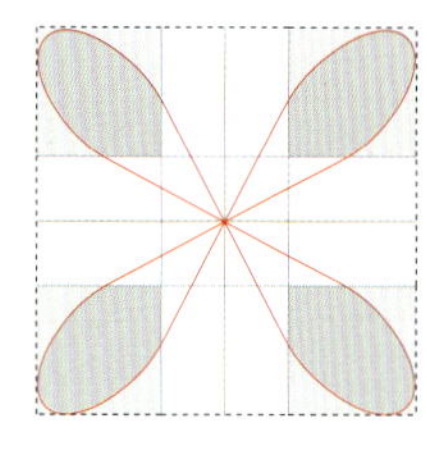

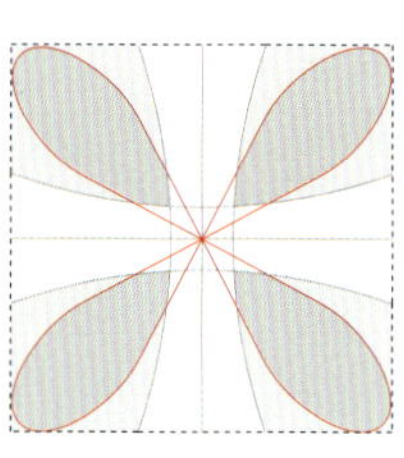

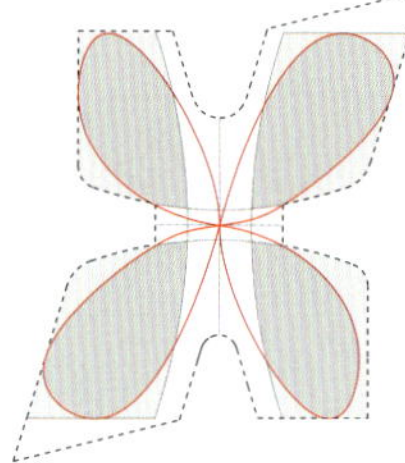

Geometry

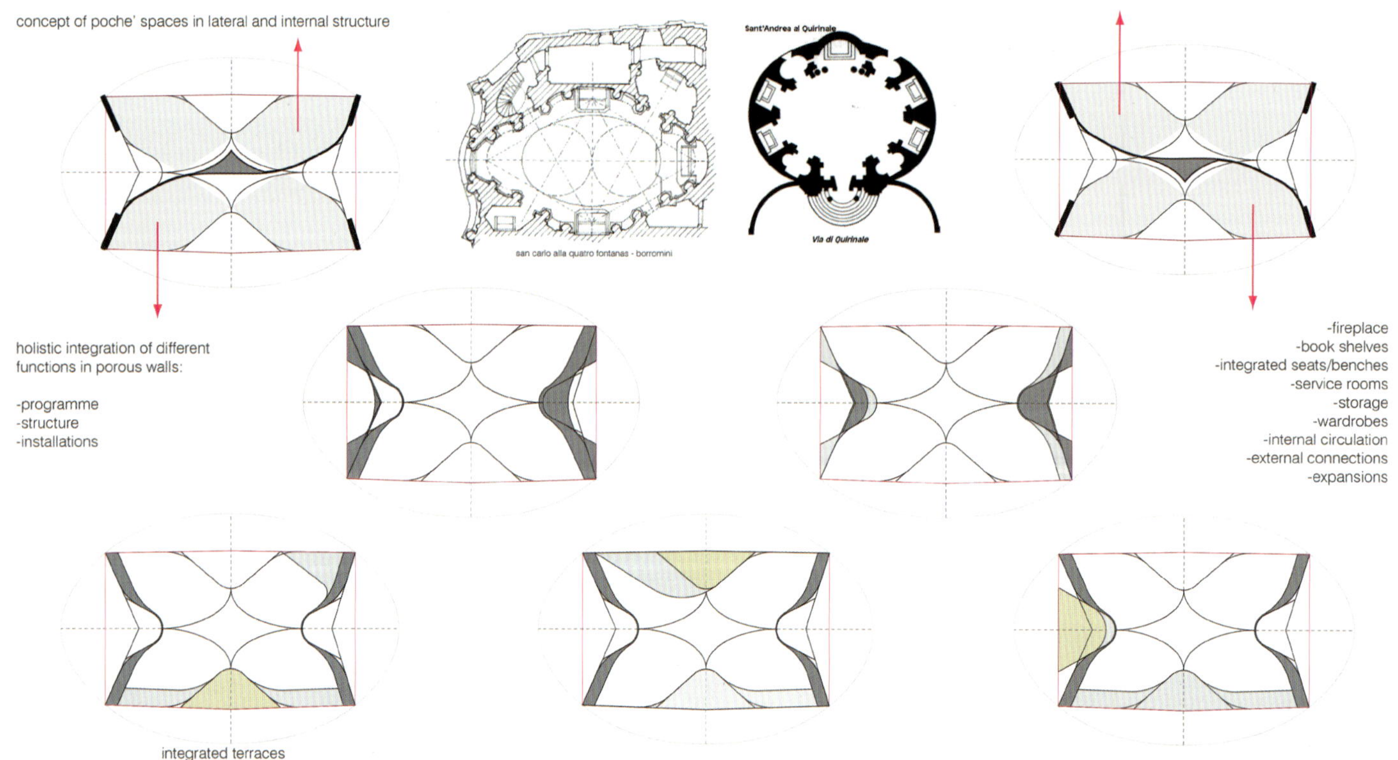

Service

© Inga Powilleit

© Inga Powilleit

© Inga Powilleit

© Fedde de Weert

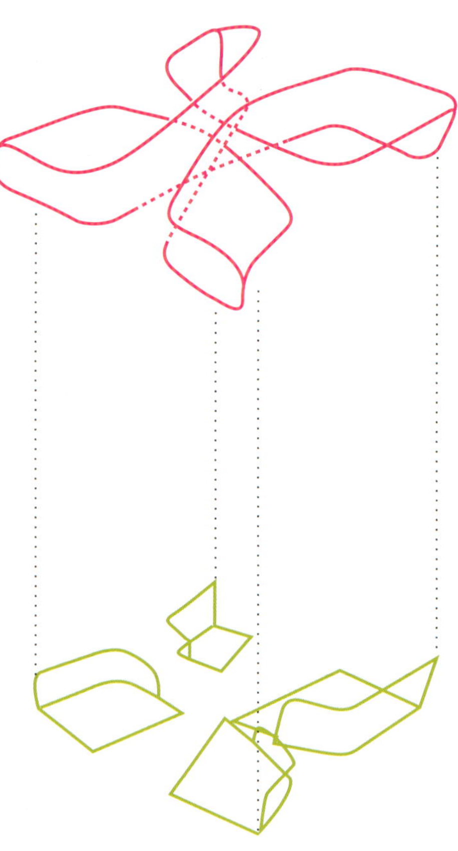

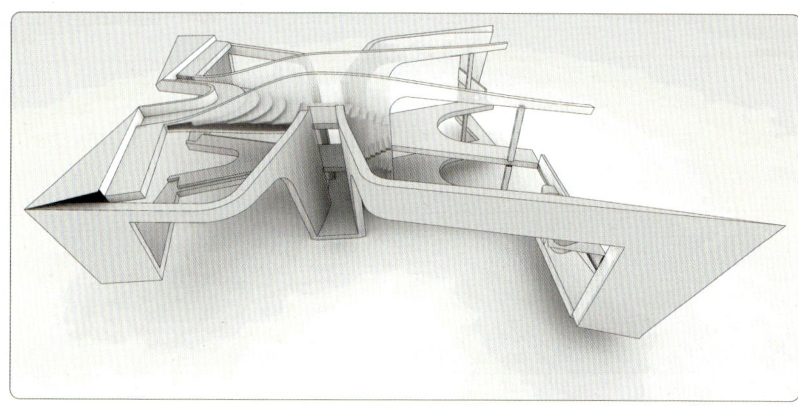

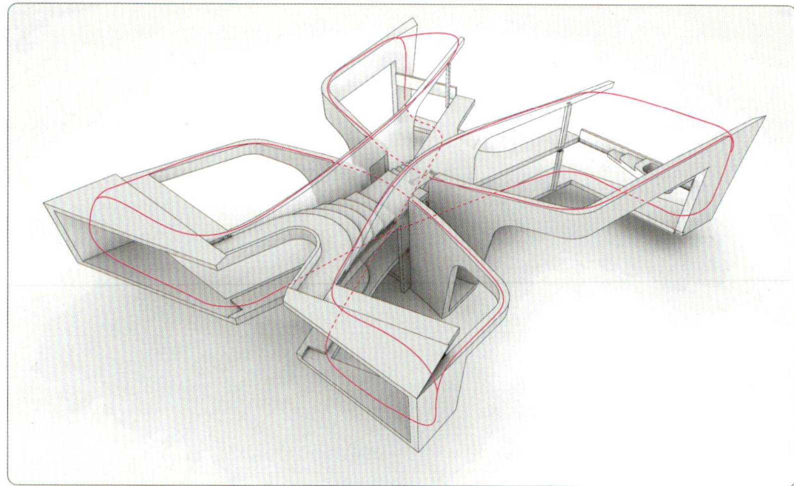

Volume Study

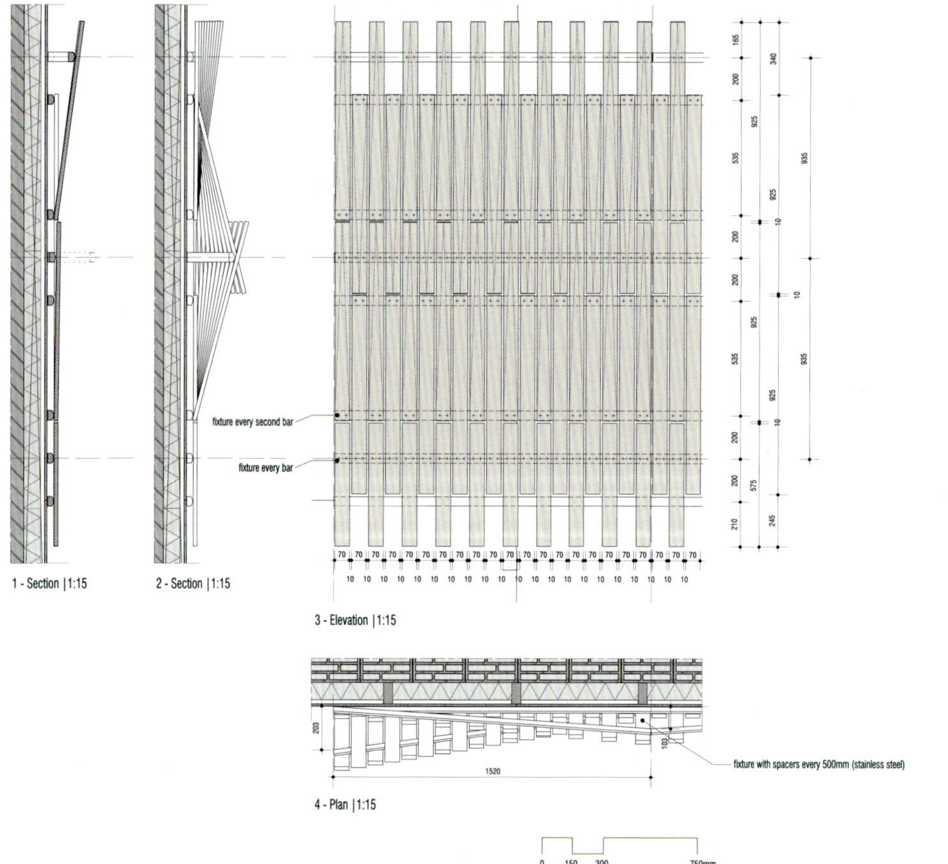

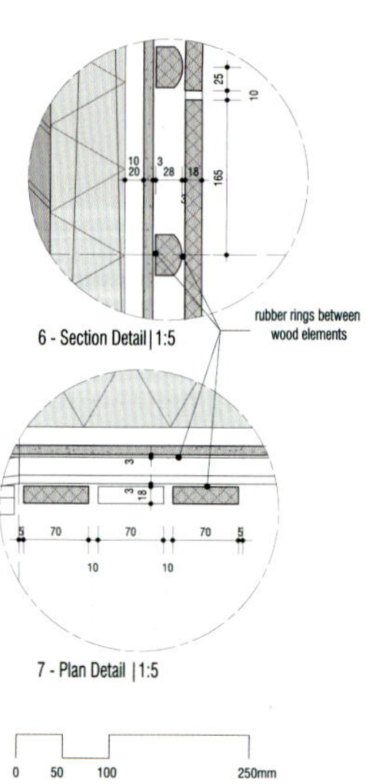

Facade Detail

© Fedde de Weert

© Fedde de Weert

Sports Center In Neudorf, Strasbourg, France

AZC

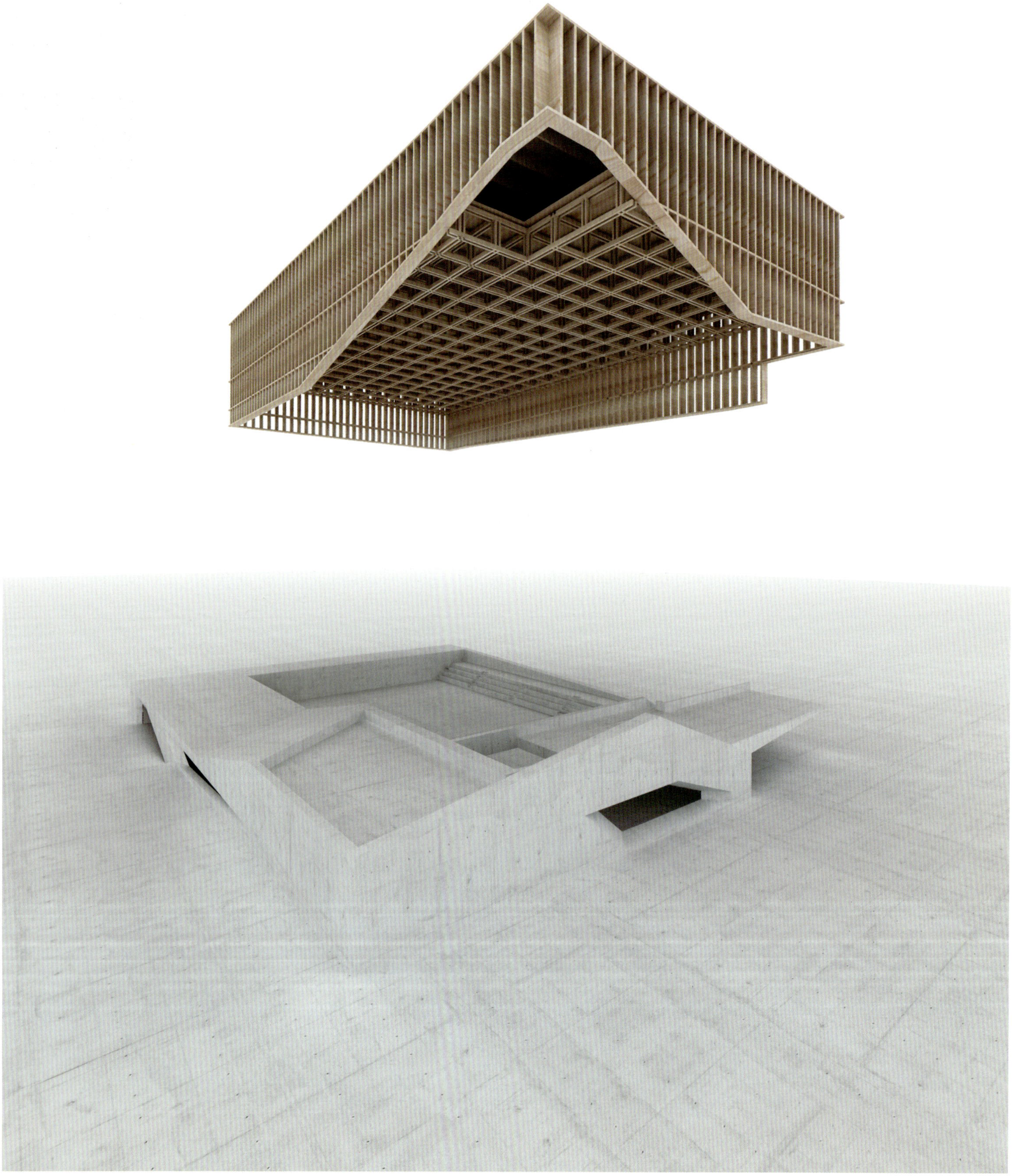

Exploded Parts Diagram

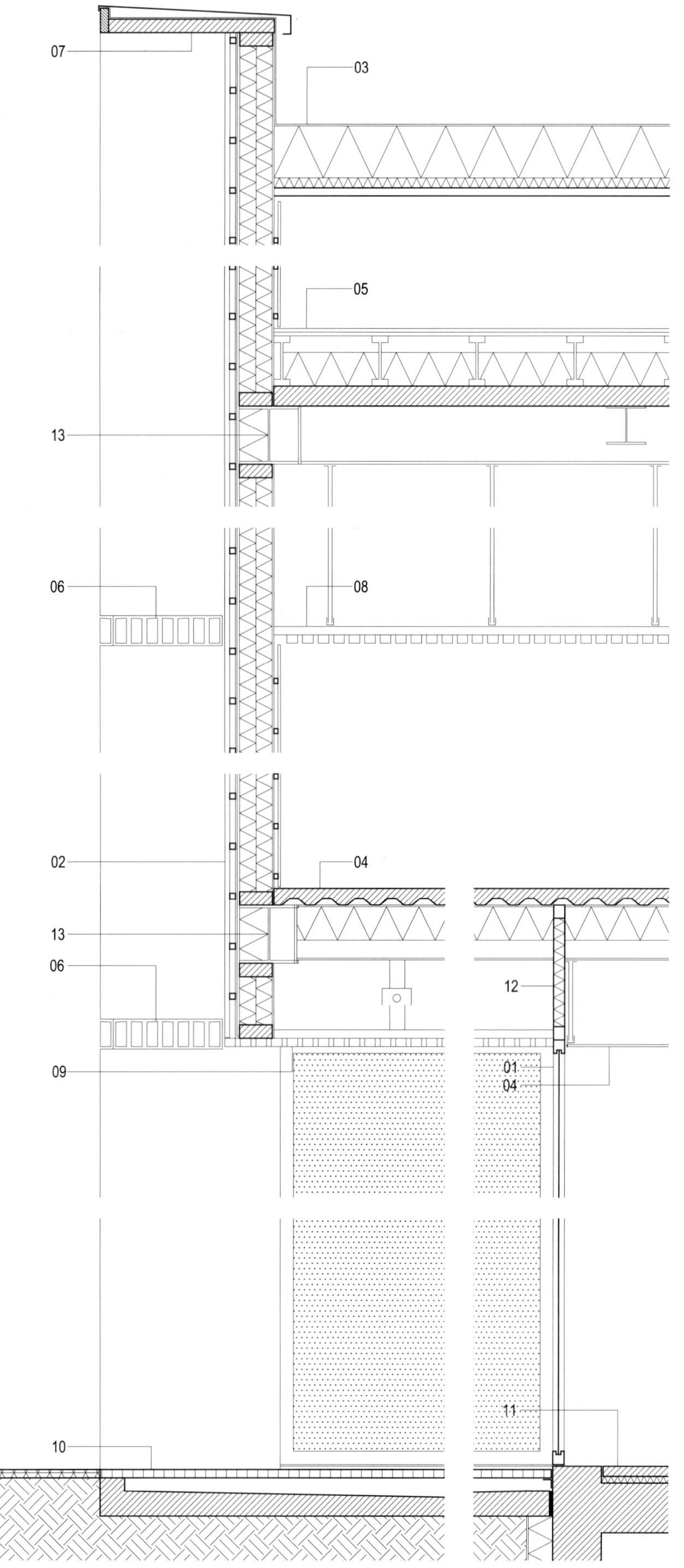

01- WINDOW WITH ALUMINIUM FRAME
02- EXTERIOR WALL:
EXTERIOR WOOD BOARDING
STONE WOOL THERMAL INSULATION
PLASTER BOARD FINISHING
03- ROOF:
ROOF SEALING
STONE WOOL THERMAL INSULATION PANELS
STONE WOOL PANEL
PERFORATED STEEL DECK SUPPORT
METAL FRAME
04- LEVEL 1 FLOOR:
COMPOSITE SLAB FLOOR
STONE WOOL THERMAL INSULATION
STEEL BEAM
05- LEVEL 2 TYPE 1 FLOOR:
METAL SHEET FLOOR FINISHING
INSULATION
WOOD PANEL
STEEL BEAM
06- HORIZONTAL WOOD BOARDING
07- WOOD PANEL, POWDER COATED ALUMINIUM SHEET
08- SUSPENDED CEILING WITH ACOUSTIC INSULATION
09- EXTERIOR WOOD BOARDING
10- STEEL DUCKBOARD
11- FLOOR HEATING WITH THERMAL INSULATION
12- INSULATION FOR WINDOW FIXING
13- STEEL I-SECTION

Detail

Longitudinal Elevation

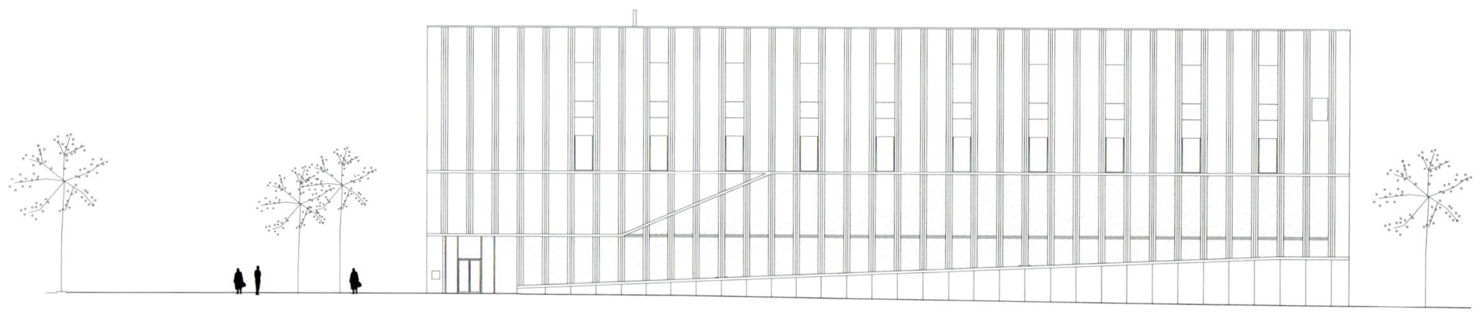

Transverse Elevation

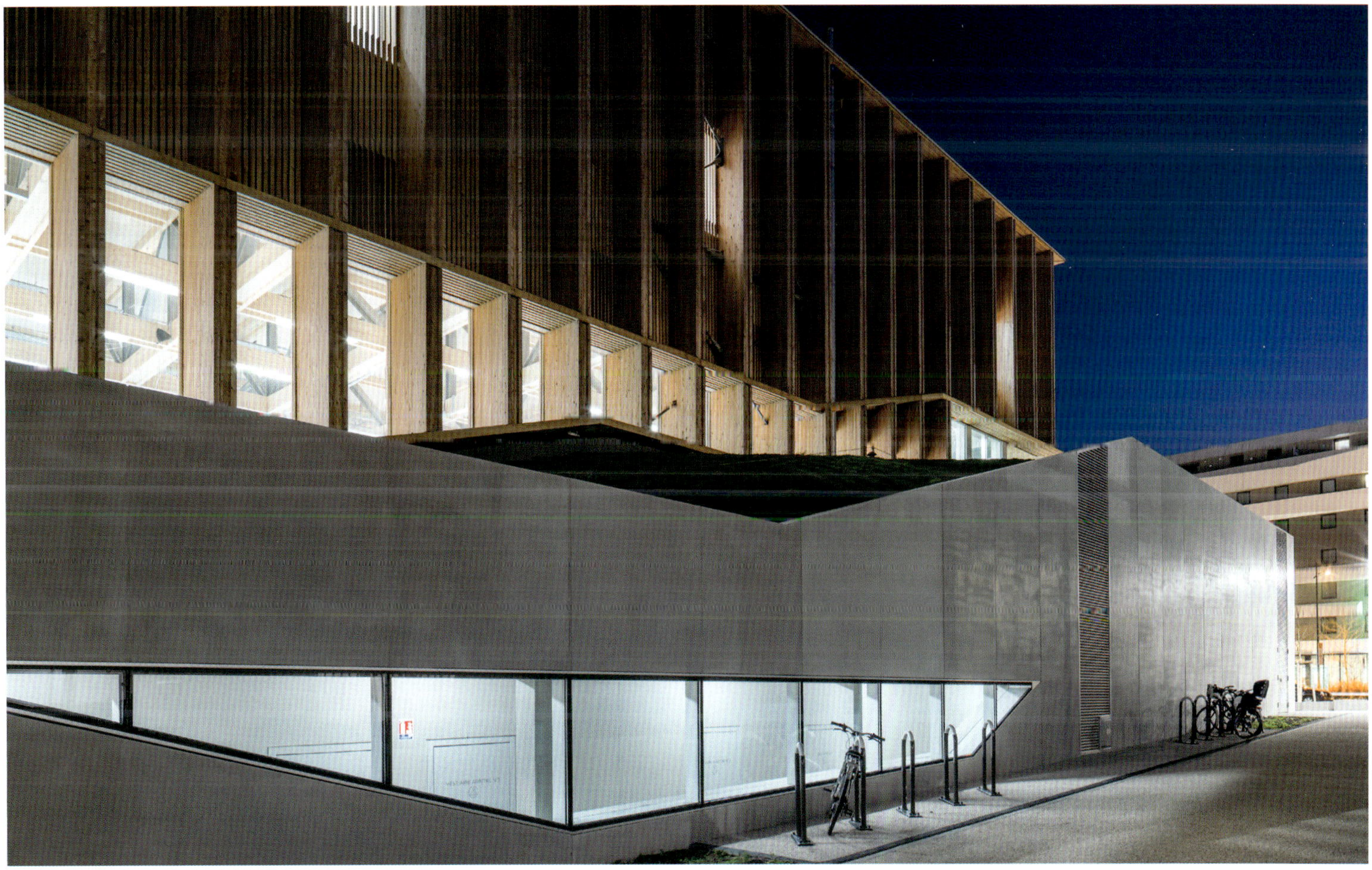

Café Grumpy, New York, USA

Z-A studio & Cheng+Snyder

© Noa Kalina

© Noa Kalina

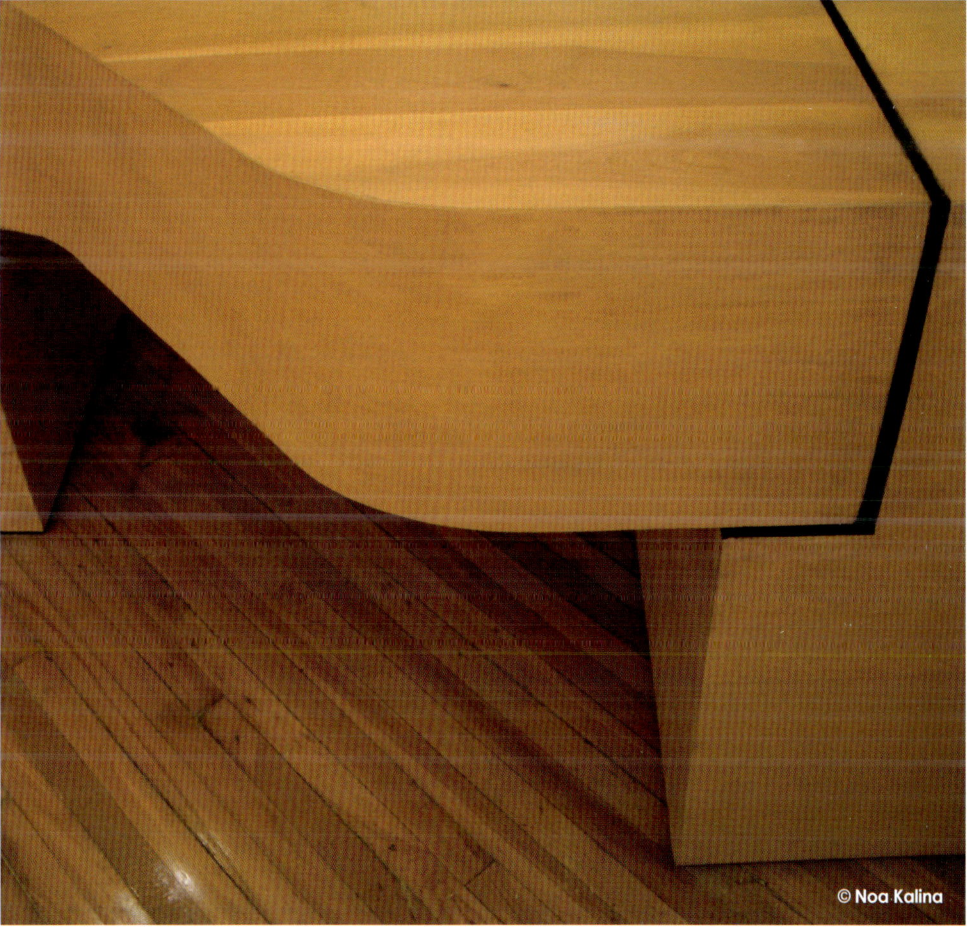
© Noa Kalina

15-Piece Coffee Bar

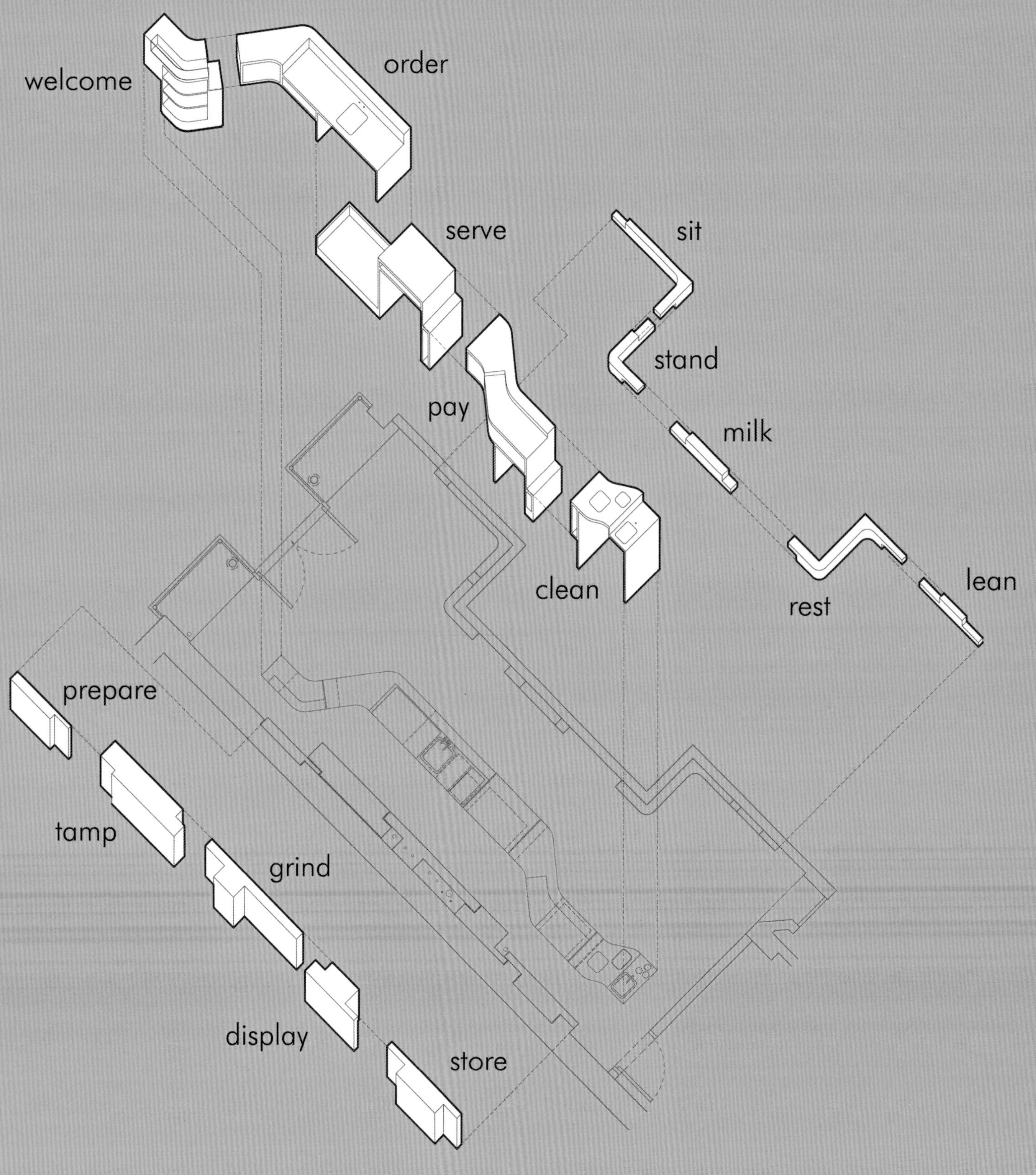

© Noa Kalina

© Noa Kalina

Oriental Clinic Bom_Flow Space, Seoul, South Korea

jay is working

© Park Wan soon

Thouugh the volume shape - Extension of sight

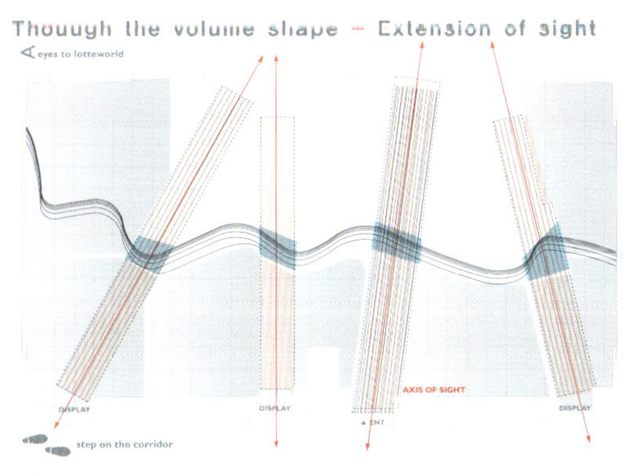

Curved line + volume shape system

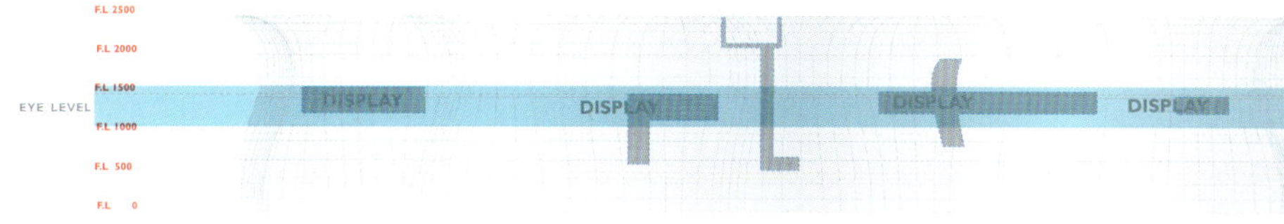

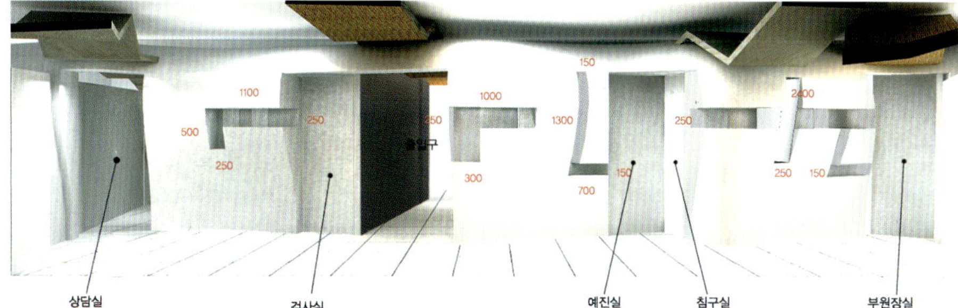

Spatial Ceiling

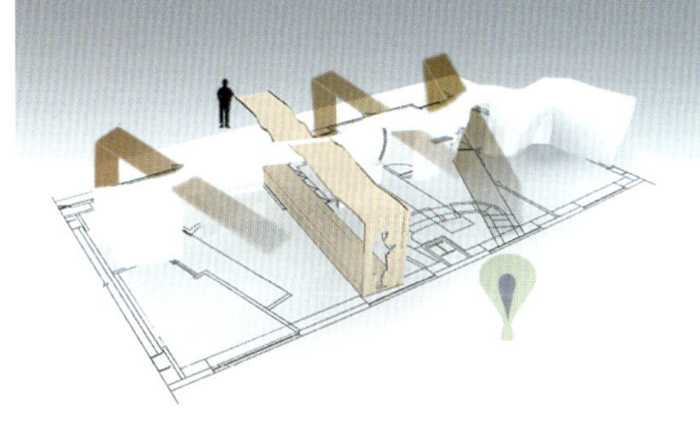

Concept

Display

Room Like, Sapporo Hokkaido, Japan

Sasaki Yusuke, Sekiguchi Satomi

OUTDOOR

Rendering

1.

2.

3.

4.

Construction Process

© Nagai Anna

© Nagai Anna

Room N3W, Sapporo Hokkaido, Japan

Sasaki Yusuke, Sekiguchi Satomi

© Nagai Anna

© Nagai Anna

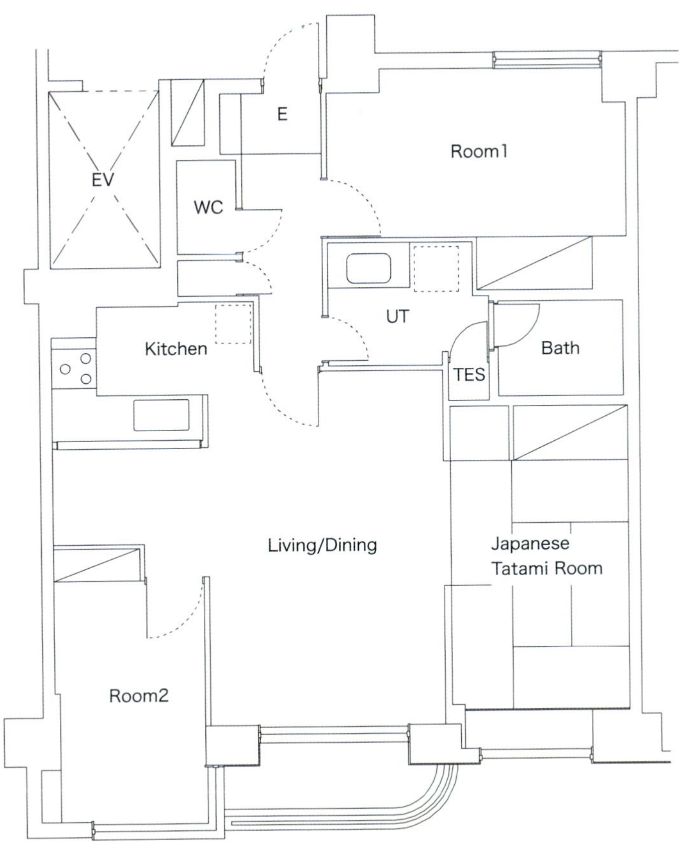

Plan (Before)

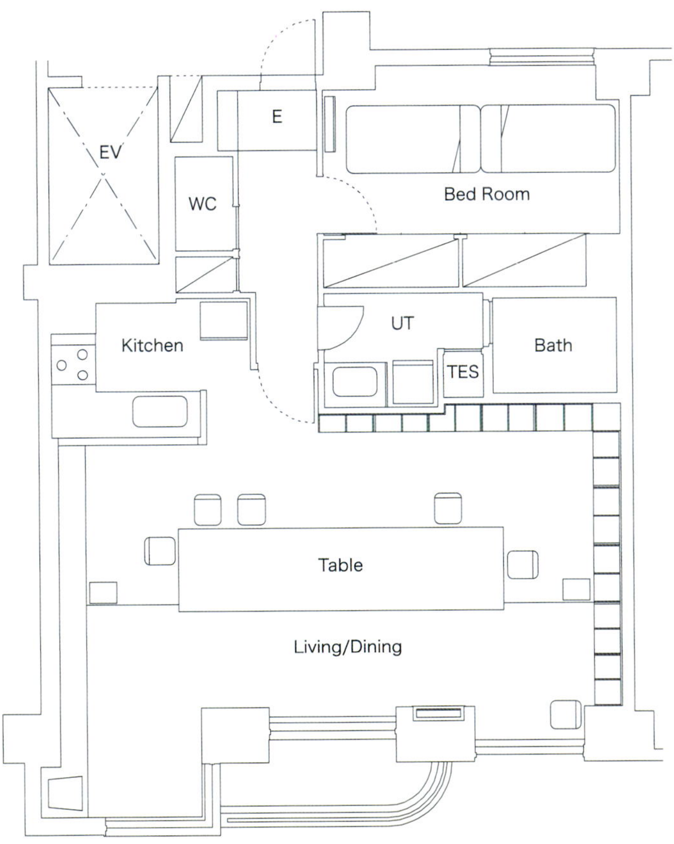

Plan (After)

Plan (Before)

Rendering

© Nagai Anna

Rosso Restaurant, Ramat Ishay, Israel
SO Architecture

© Asaf Oren

Concept / Ceiling elments

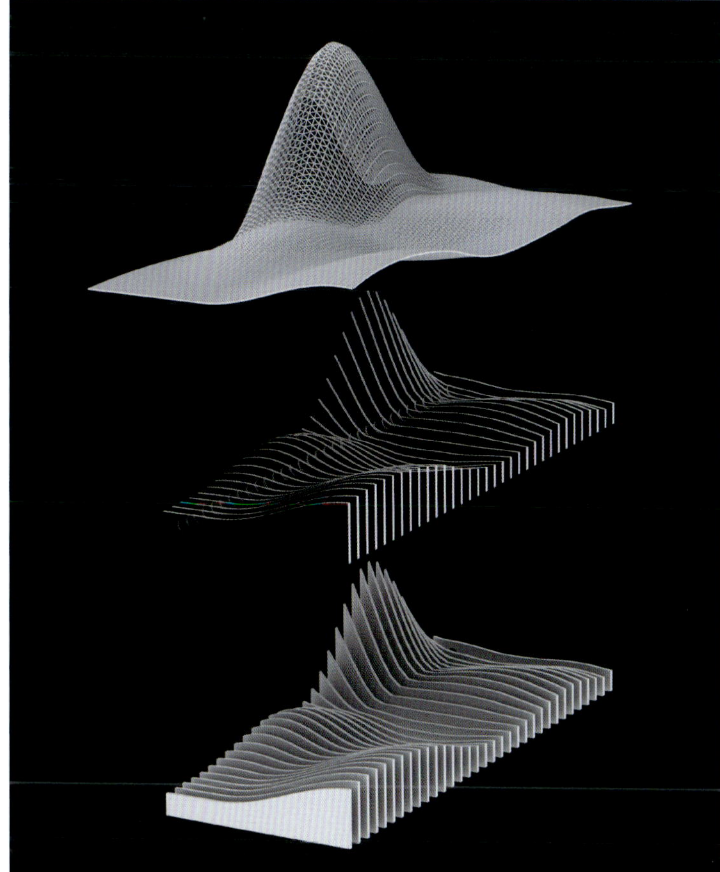

Ceiling Computer Model

© Asaf Oren

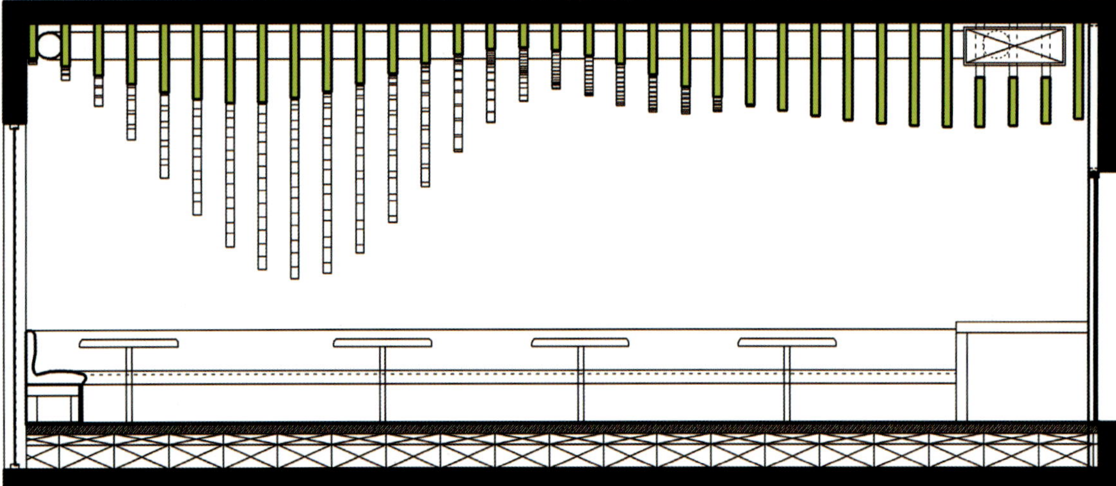

Section

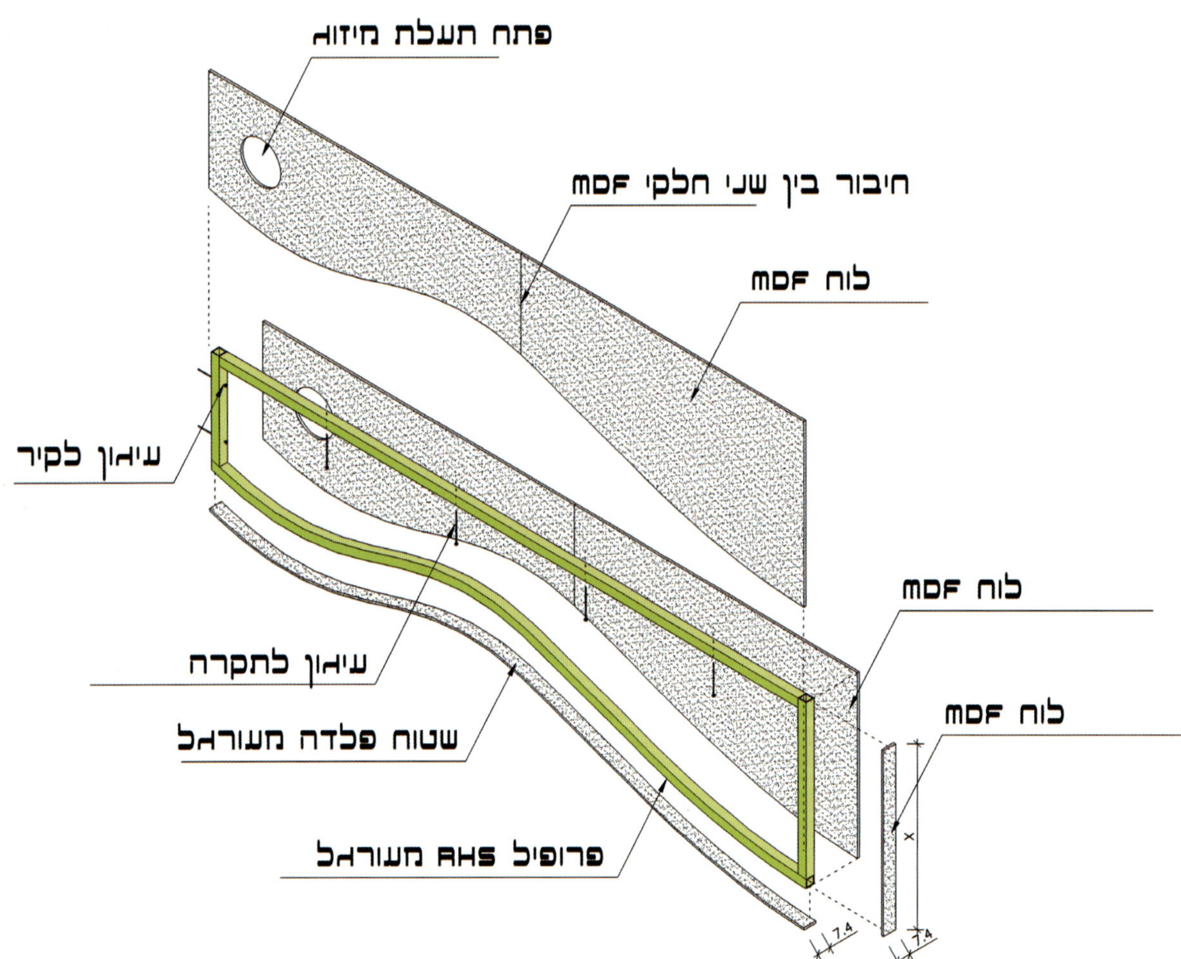

Detail

© Asaf Oren

Urbia Furniture System for Small Apartments in Big Cities, New York, USA

OBRA Architects

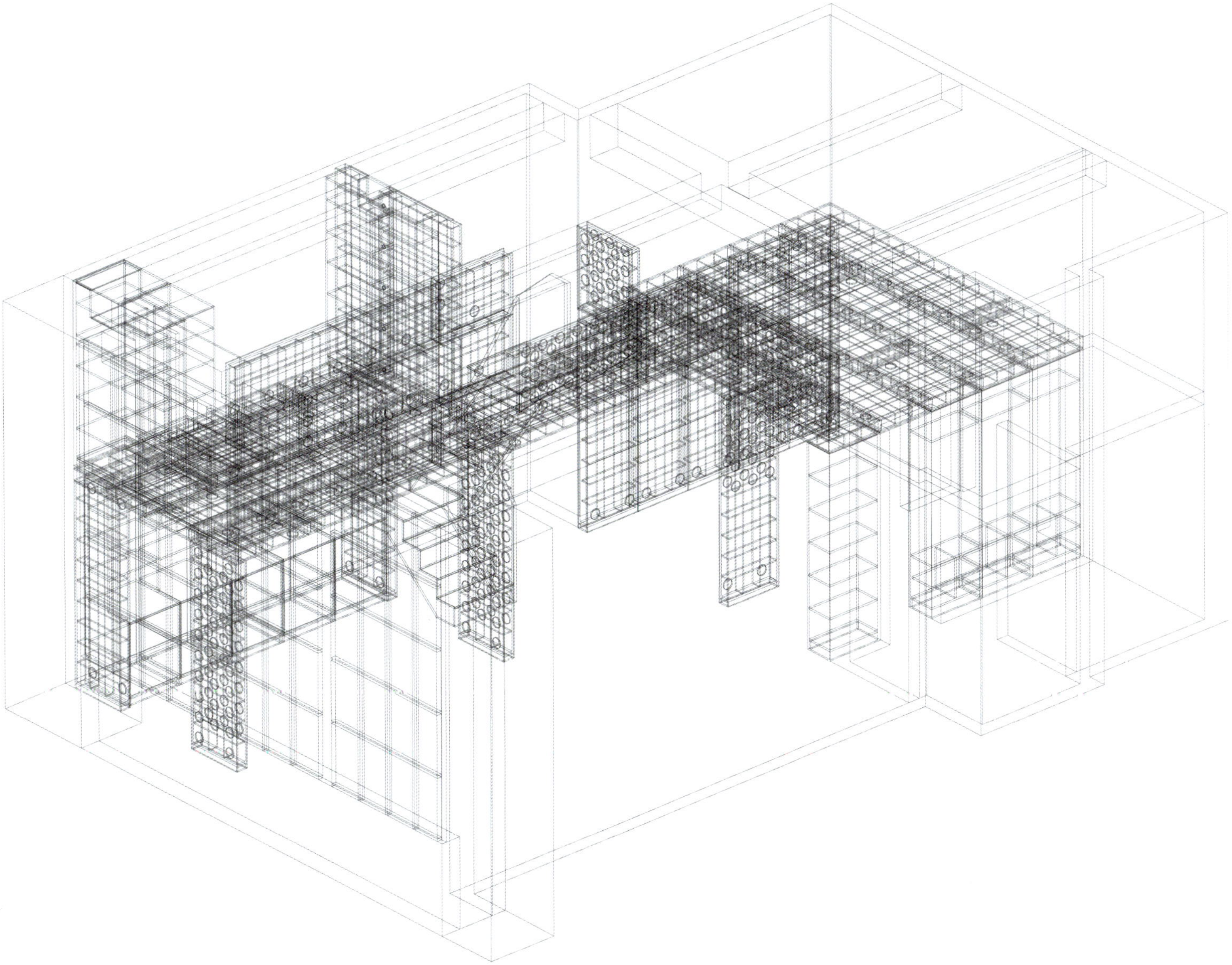

Liine Drawing

Construction Detail

1.

2.

3.

5.

6.

7.

9.

10.

11.

13.

14.

15.

17.

18.

19.

Construction Process

4.

8.

12.

16.

20.

flip, Furniture

marco hemmerling architecture design

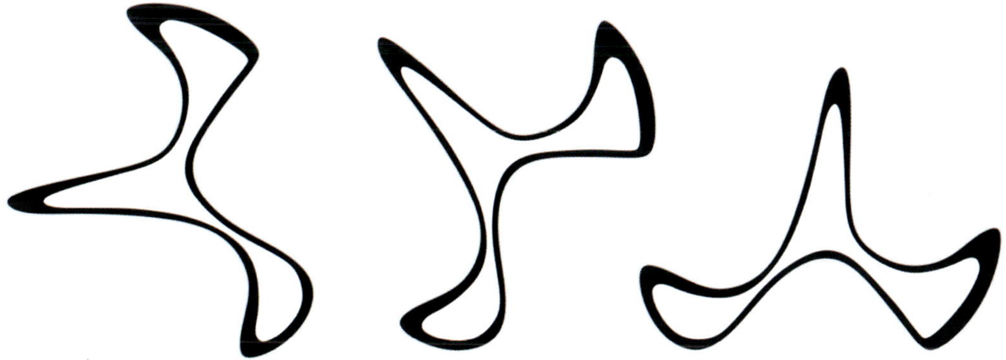

A house in Awajishima, Hyogo, Japan

Dai Nagasaka

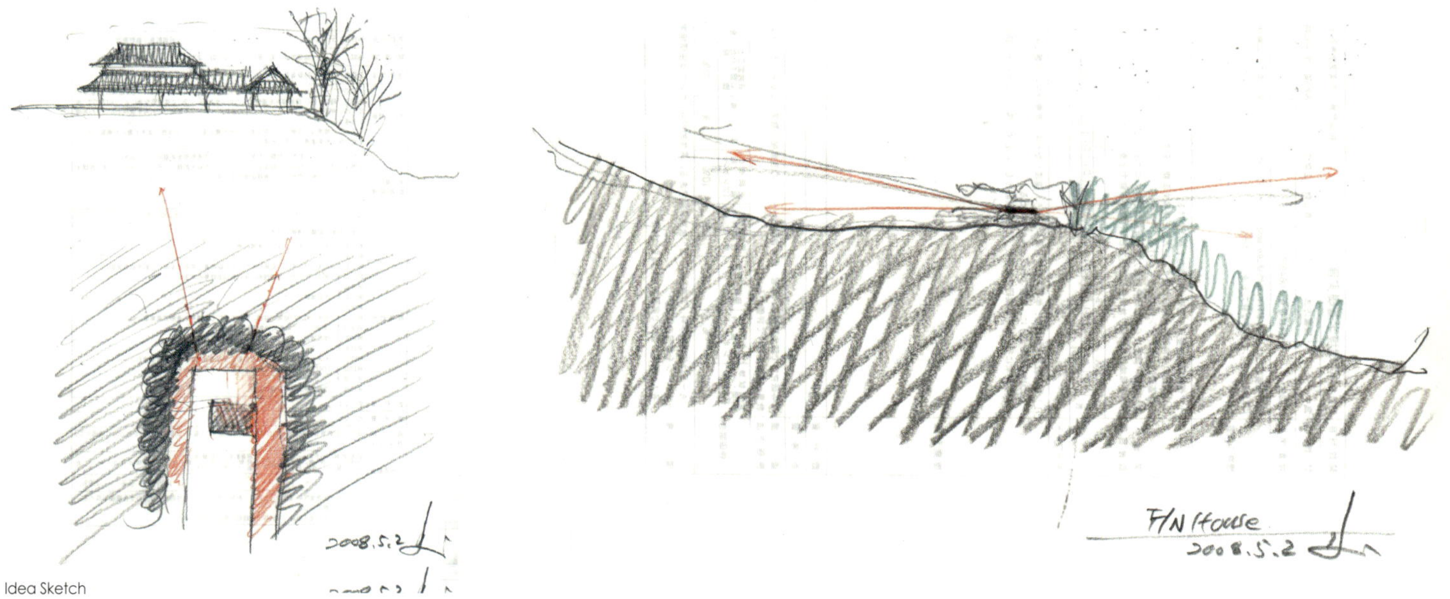

Idea Sketch

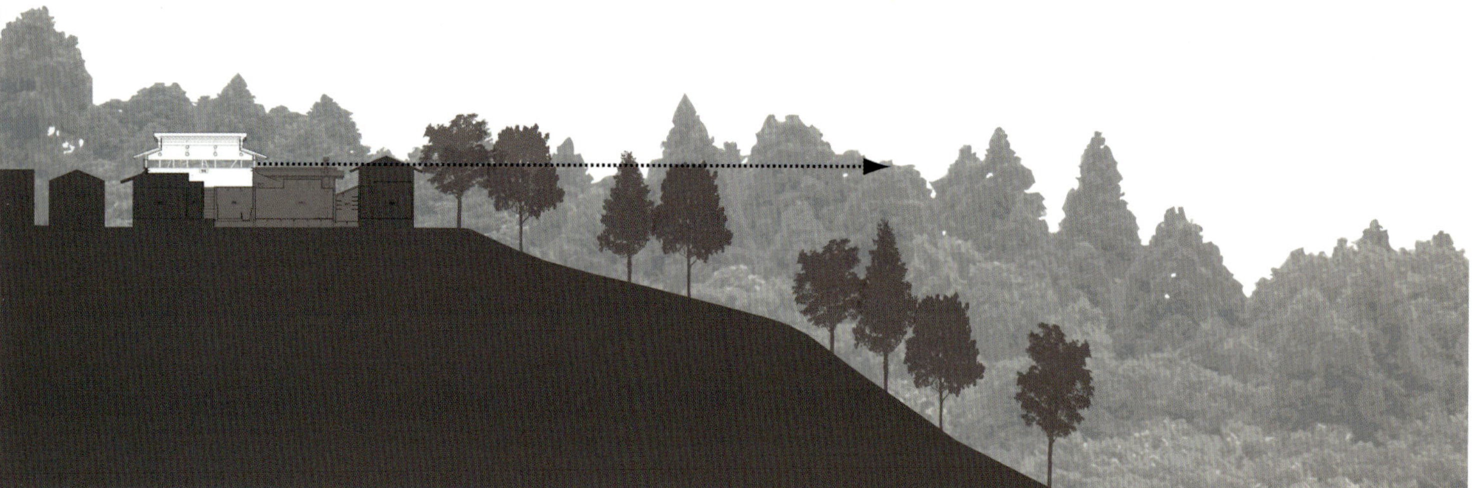
View in the Place

ugino

© Kei Sugino

© Kei Sugino

A house in Tojiin2, Kyoto, Japan

Dai Nagasaka

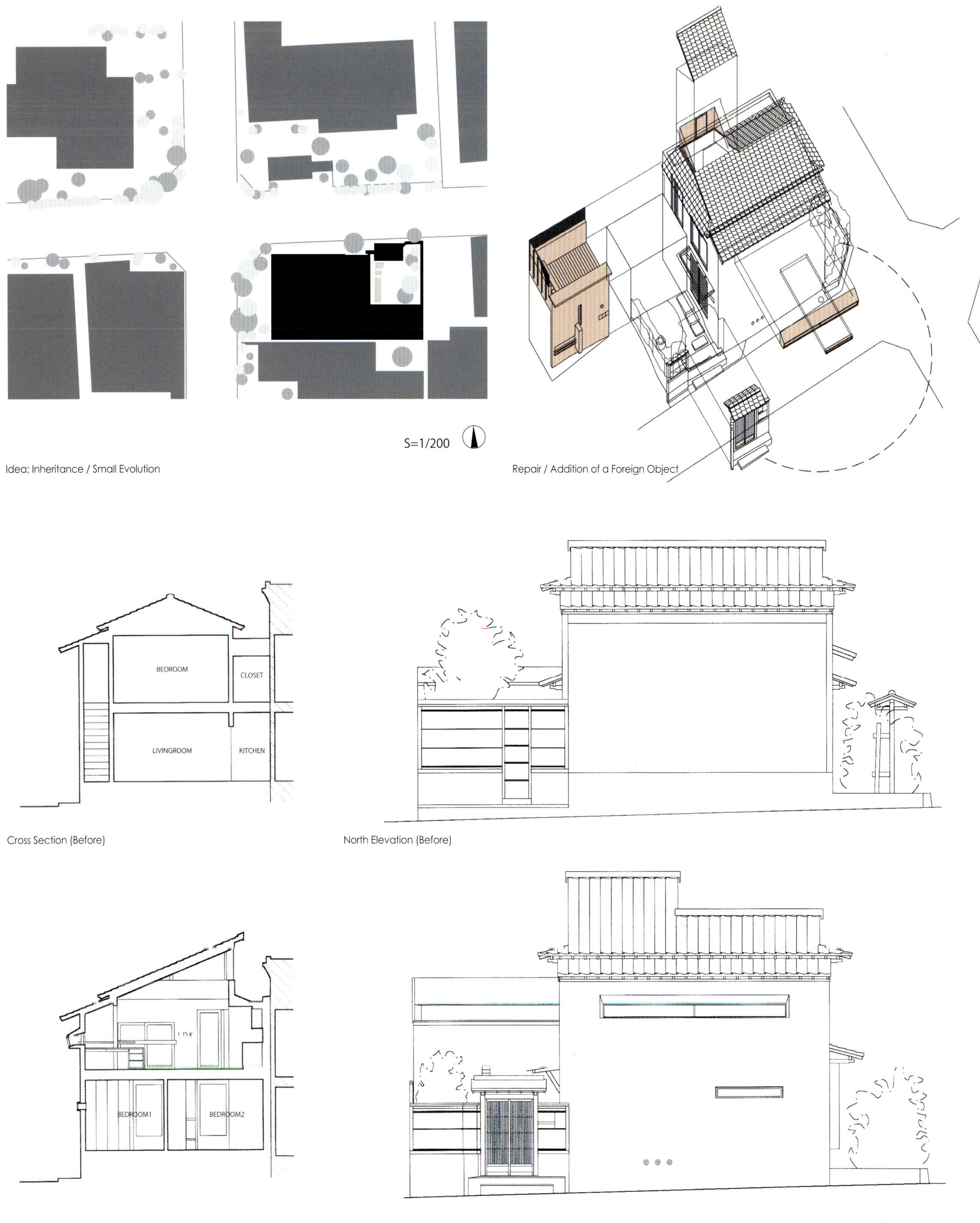

Idea: Inheritance / Small Evolution

Repair / Addition of a Foreign Object

Cross Section (Before)

North Elevation (Before)

Cross Section (After)

North Elevation (After)

Before

1.

2.

3.

4.

5.

Construction Process

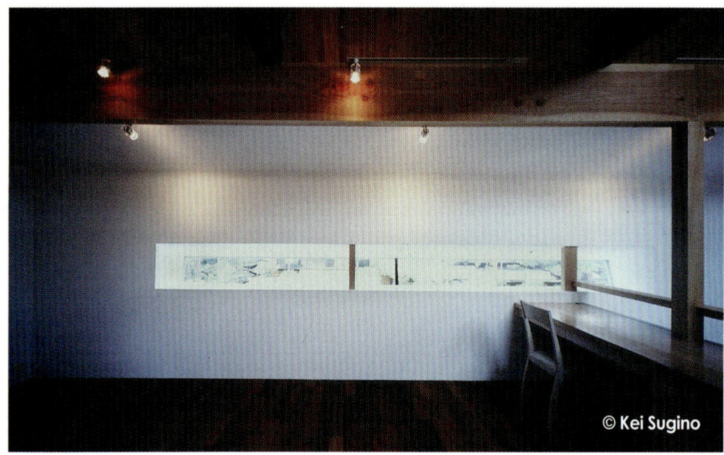
© Kei Sugino

© Kei Sugino

© Kei Sugino

© Kei Sugino

Barrow House, Melbourne, Australia

ANDREW MAYNARD ARCHITECTS

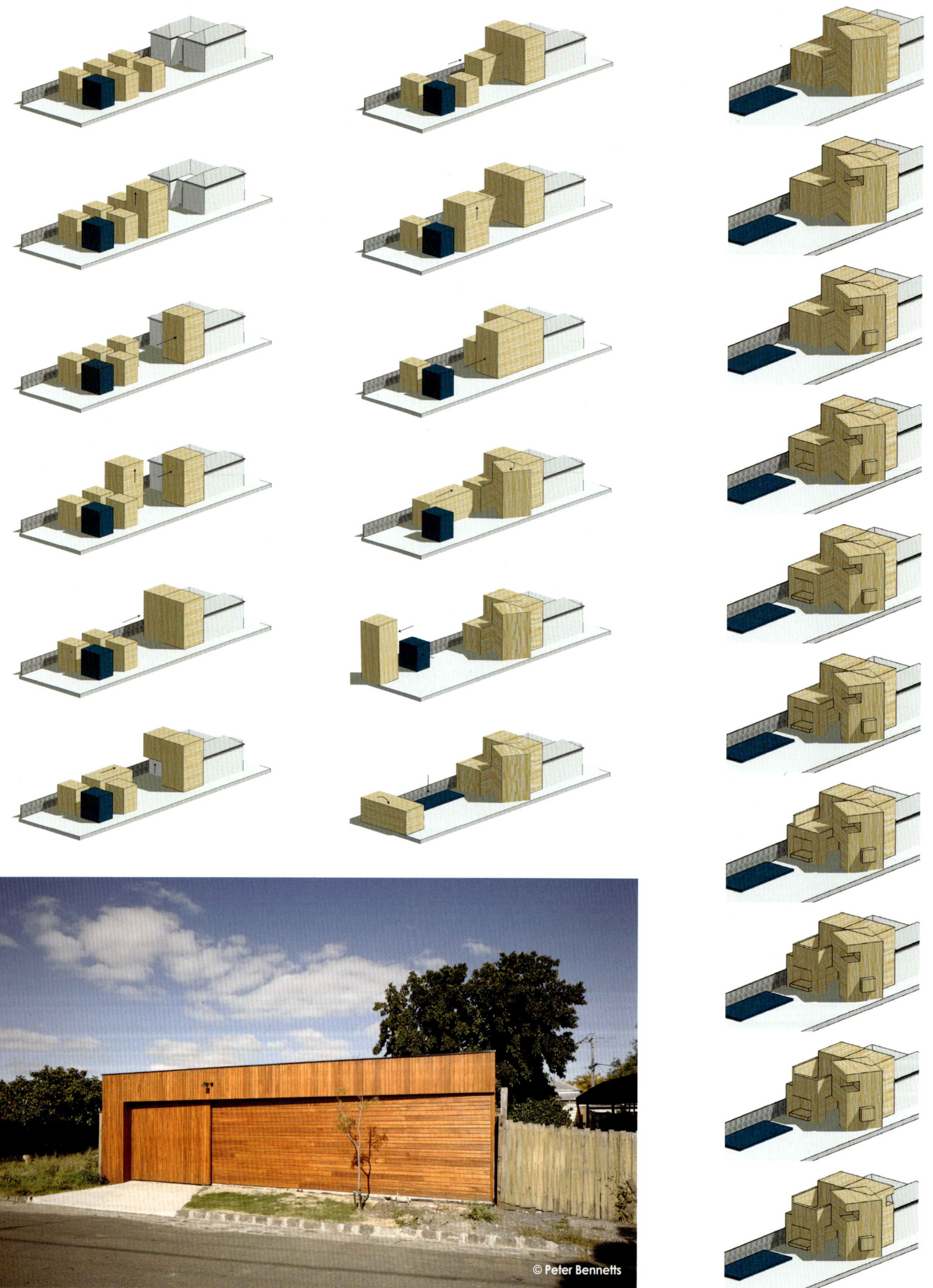
© Peter Bennetts

© Peter Bennetts

© Peter Bennetts

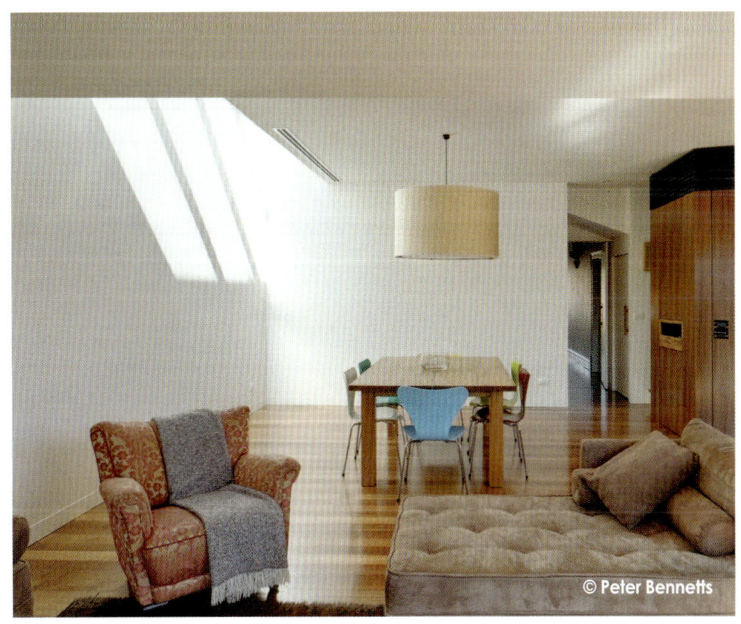

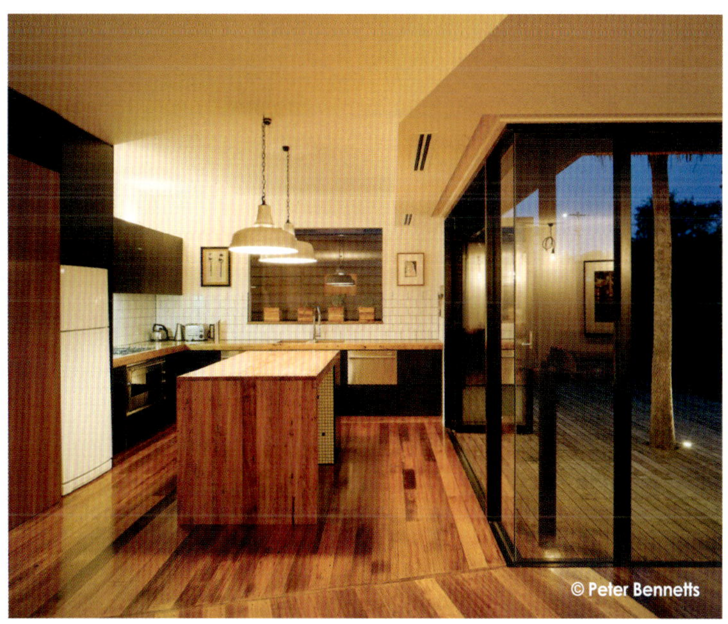

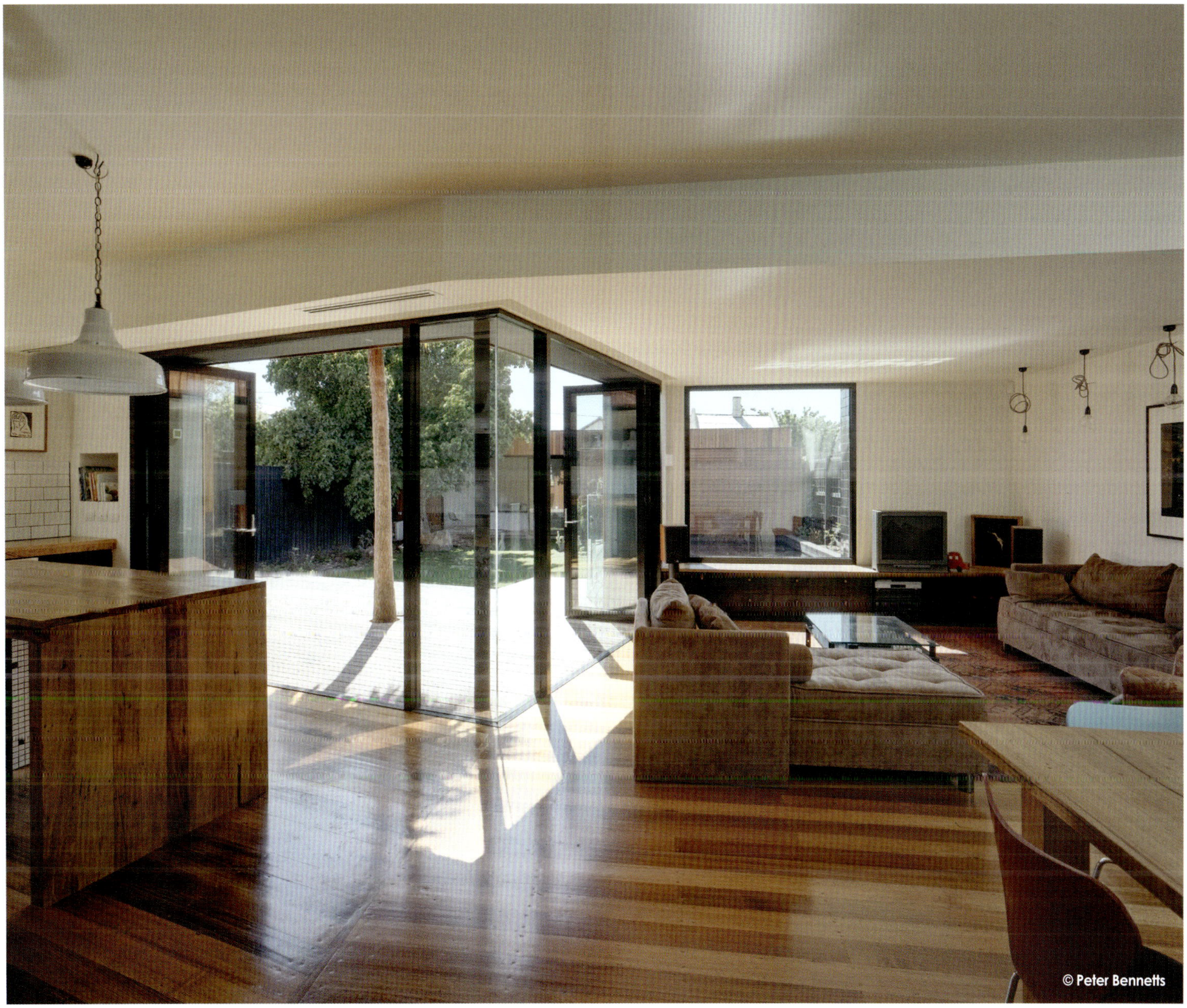
© Peter Bennetts

© Peter Bennetts

© Peter Bennetts

© Peter Bennetts

Basic House, Hyogo, Japan

Keiichi Hayashi Architect

© Yoshiyuki Hirai

Doma

Ground Floor

Living space

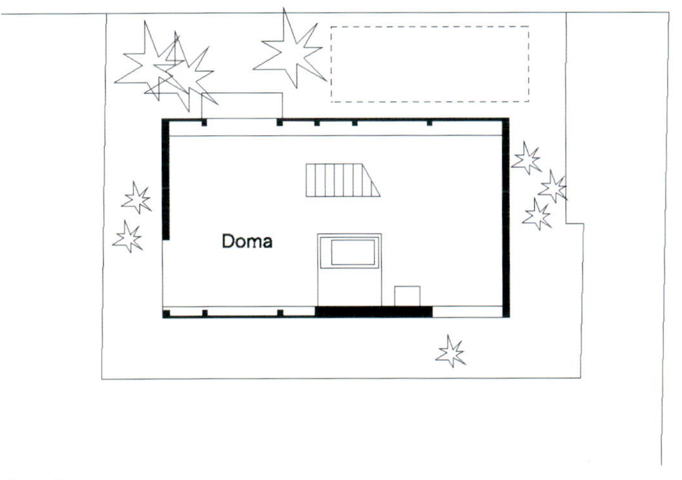

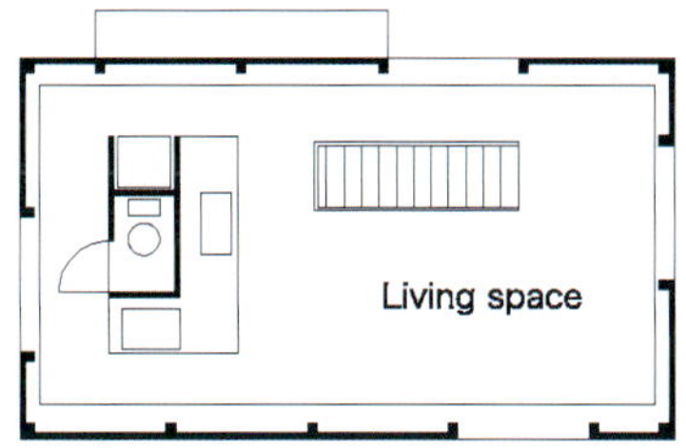

First Floor

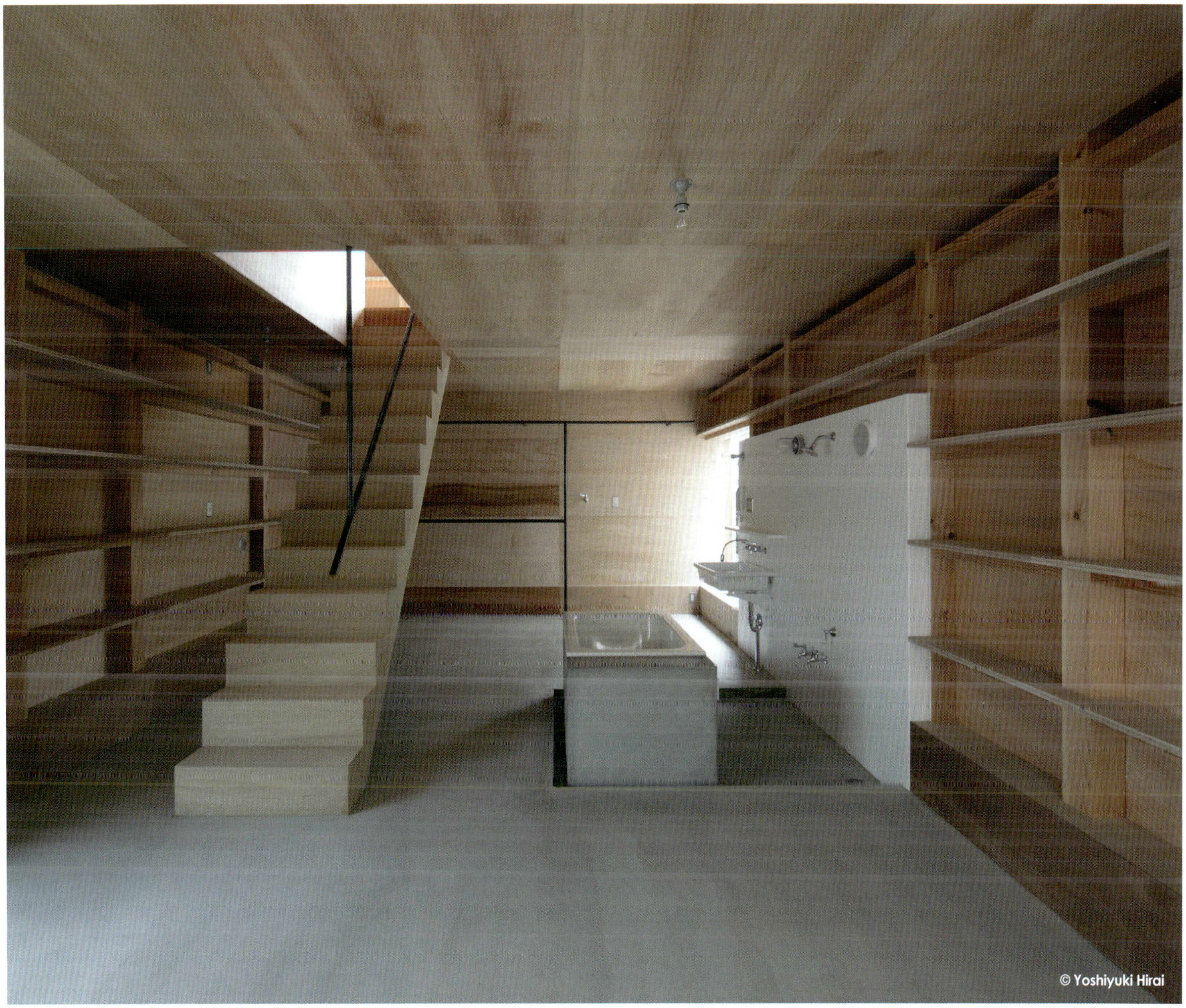

© Yoshiyuki Hirai

© Yoshiyuki Hirai

© Yoshiyuki Hirai

© Yoshiyuki Hirai

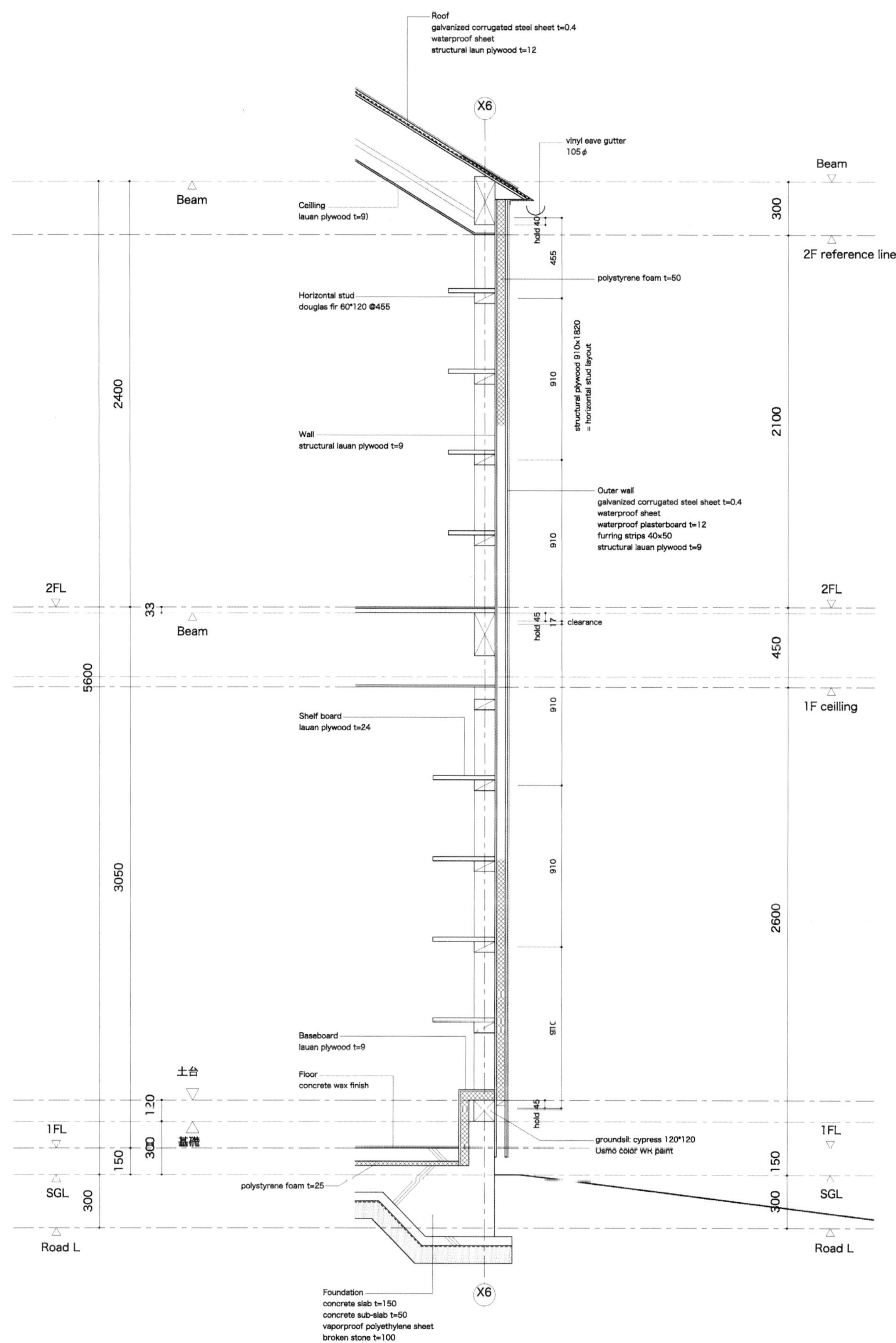

Section Detail

© Yoshiyuki Hirai

© Yoshiyuki Hirai

© Yoshiyuki Hirai

Blooming Bamboo Home, Tu Liem, Ha Noi
H&P Architects

1st FLOORPLAN

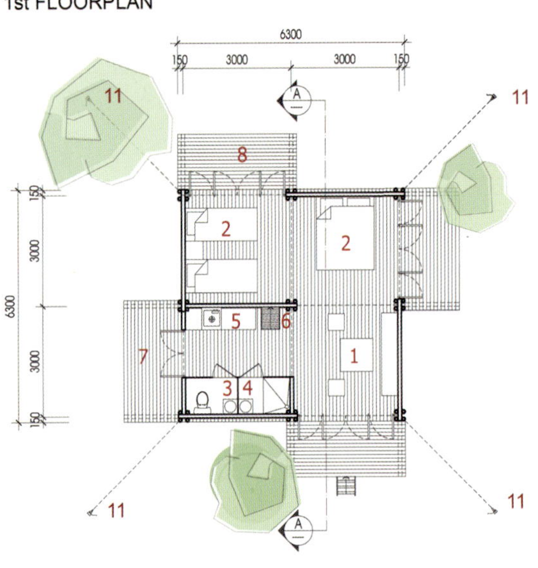

1. Living room
2. Bedroom
3. WC
4. Bathroom
5. Kitchen
6. Stair
7. Laundry + Drying
8. Outdoor terrace
9. Indoor terrace (sleeping + learning + worship)
10. area breed animal / plant
11. Anchor steel
12. Rain water tank
13. Clean water tank (filtered)
14. Waste water tank
15. Water for gardening
16. Discharged to (after treated)
17. Filter tank for rain water
18. Rain water cleaned and returned to the environment (underground reloading)

0 3M

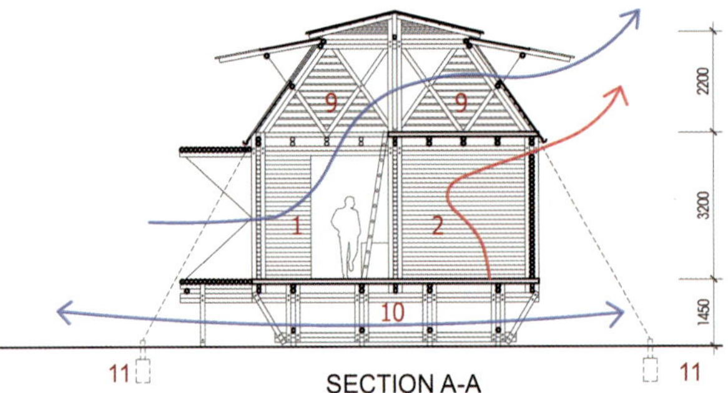

SECTION A-A

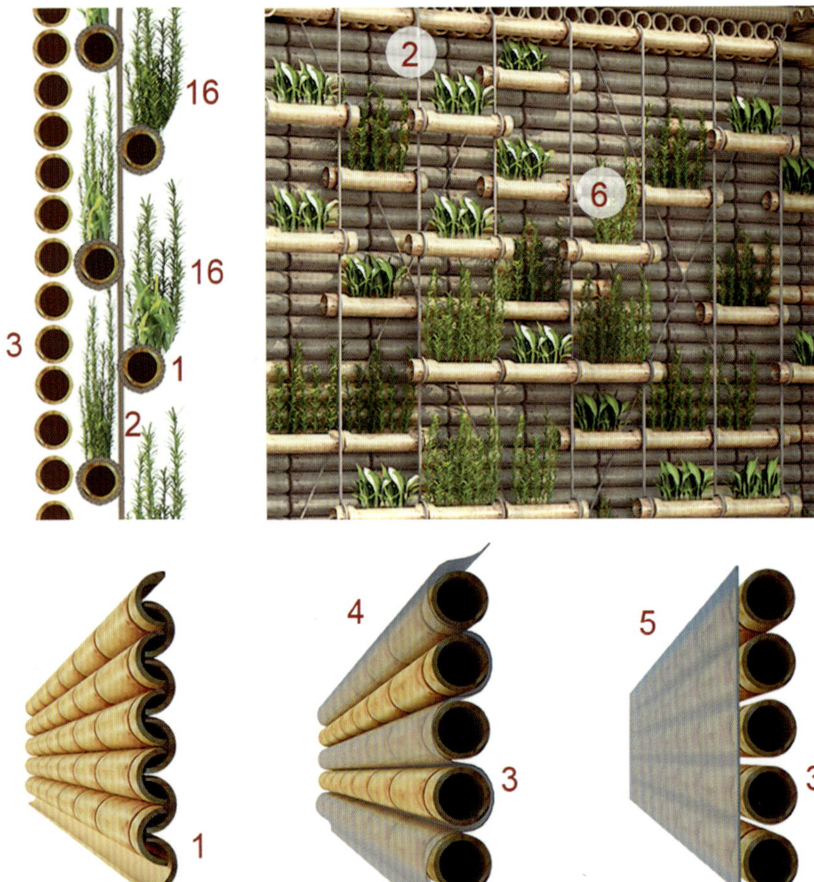

Wall Materials

1. Bamboo of 8-10cm diameter
2. Rope
3. Bamboo of 4-5cm diameter
4. Nylon Sheet(rain shield)
5. Polycarbonate Sheet
6. Vertical Garden(Vegetable, Plant, Flower...

© Doan Thanh Ha

© Doan Thanh Ha

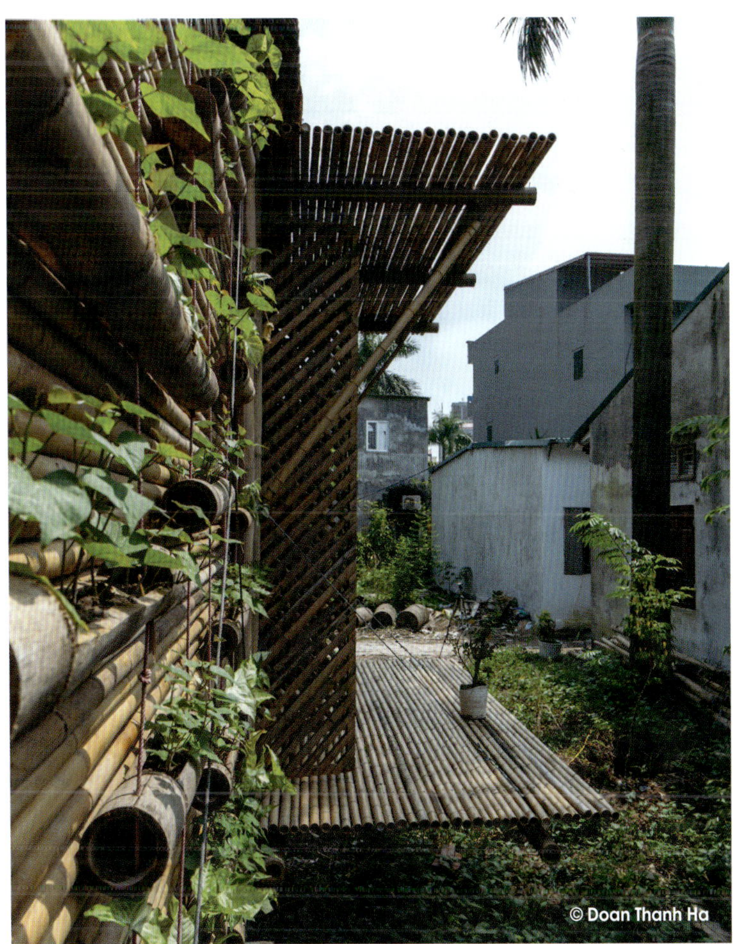

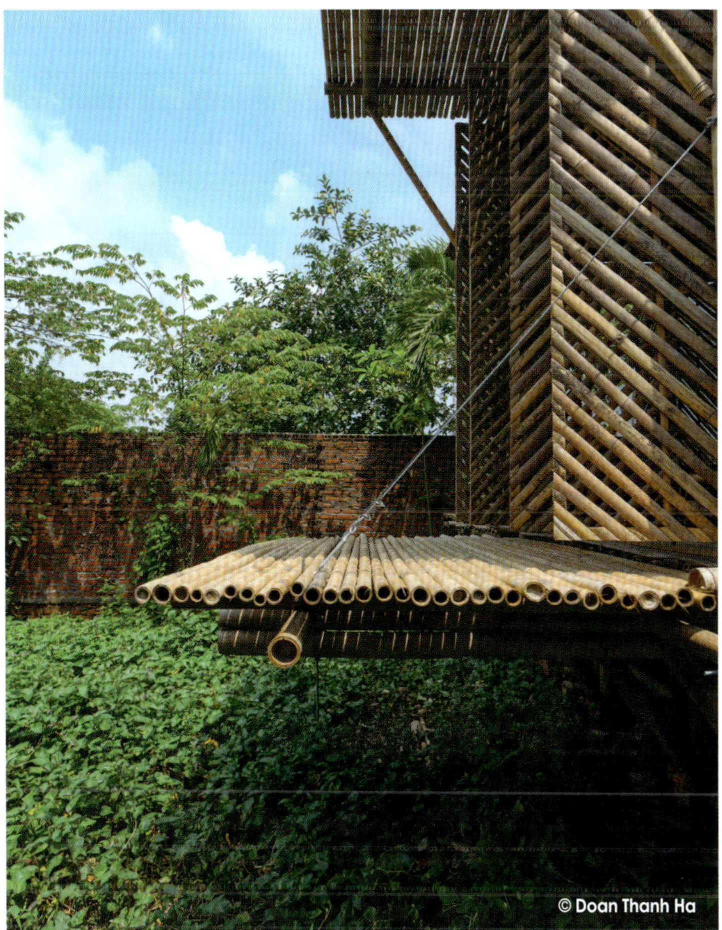

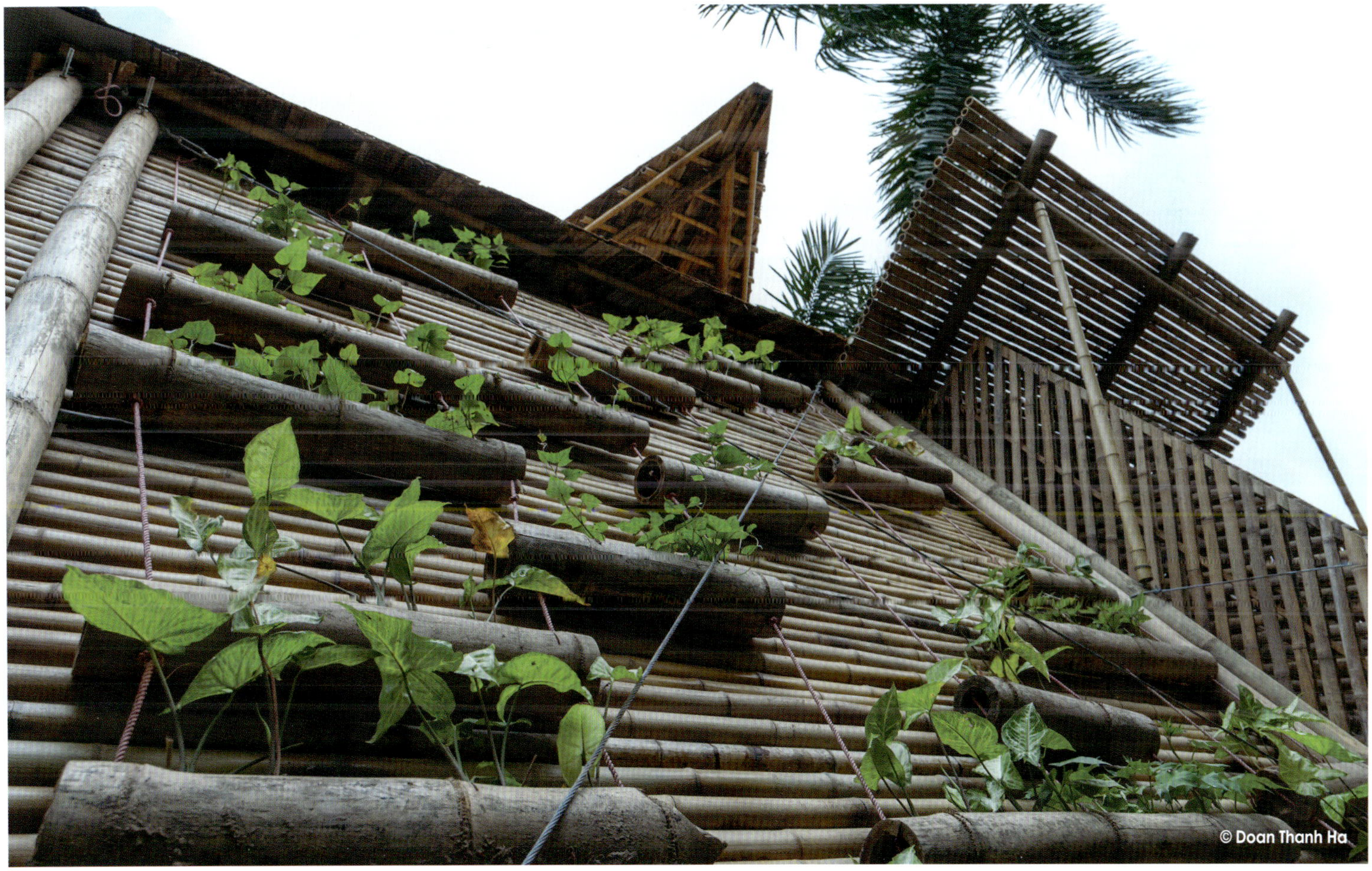
© Doan Thanh Ha

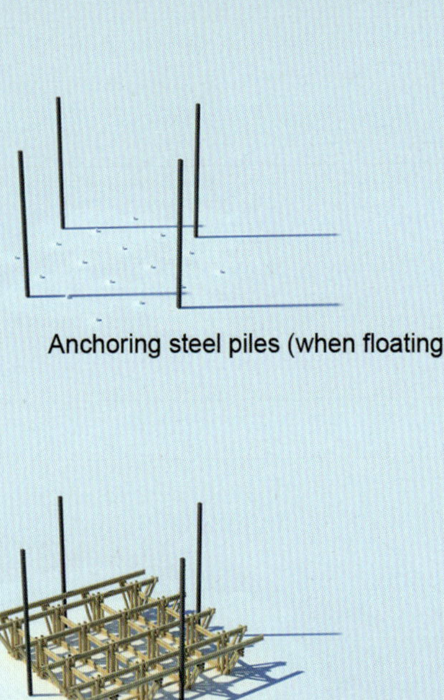

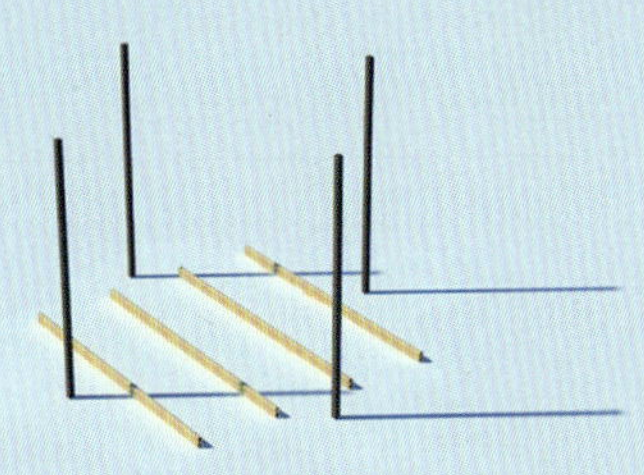
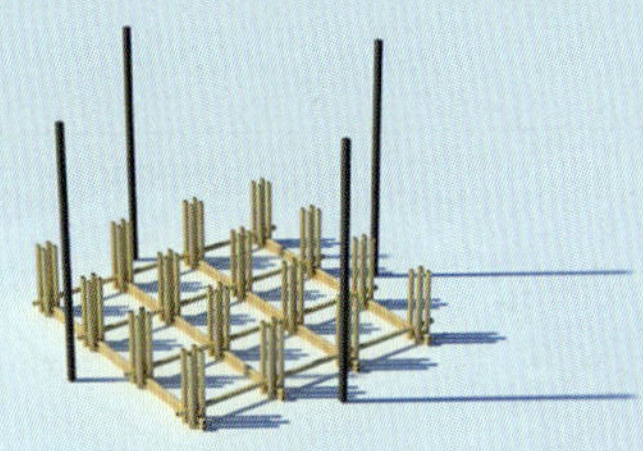
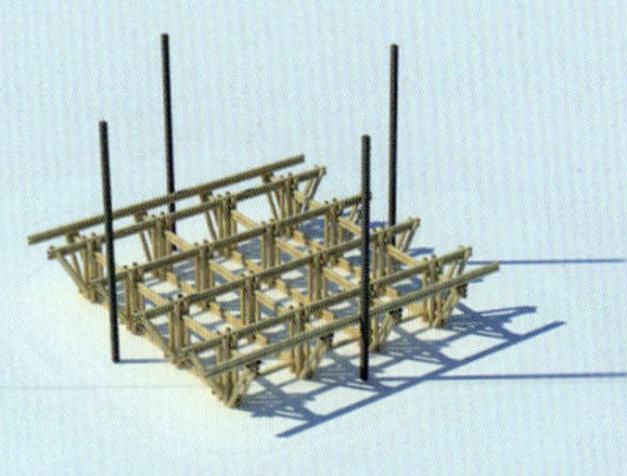

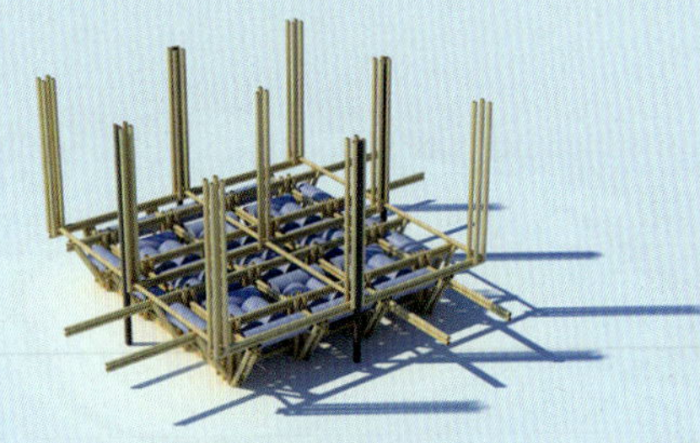

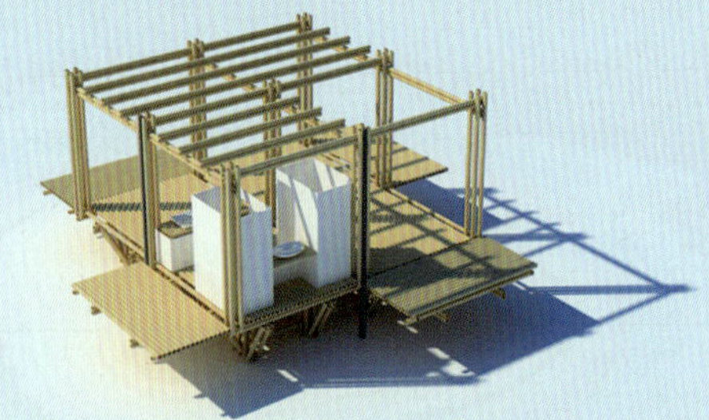

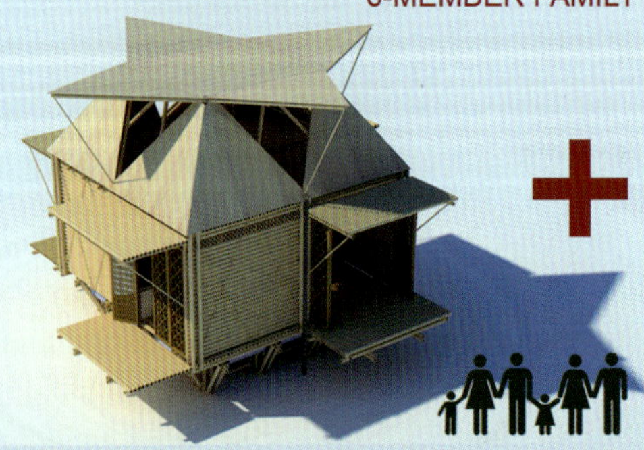

Building Process

© Doan Thanh Ha

© Doan Thanh Ha

© Doan Thanh Ha

© Doan Thanh Ha

© Doan Thanh Ha

Casa Granada, São Paulo, Brazil
Estúdio HAA!

Study Modeling

Perspective CG

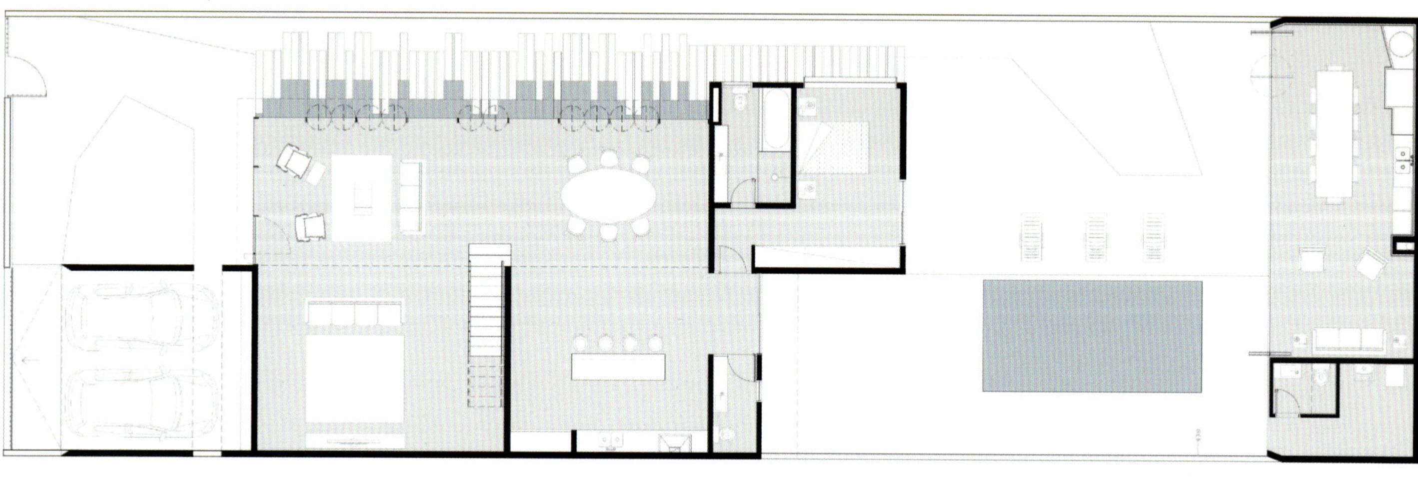

Ground Floor

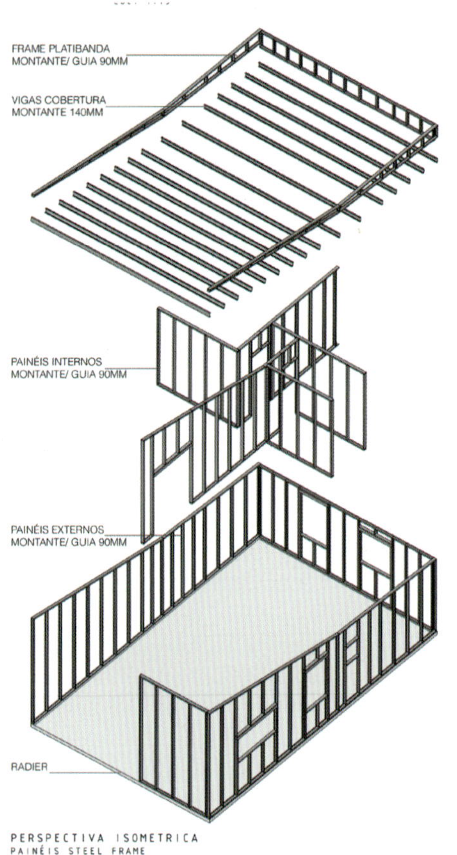

Steel Frame

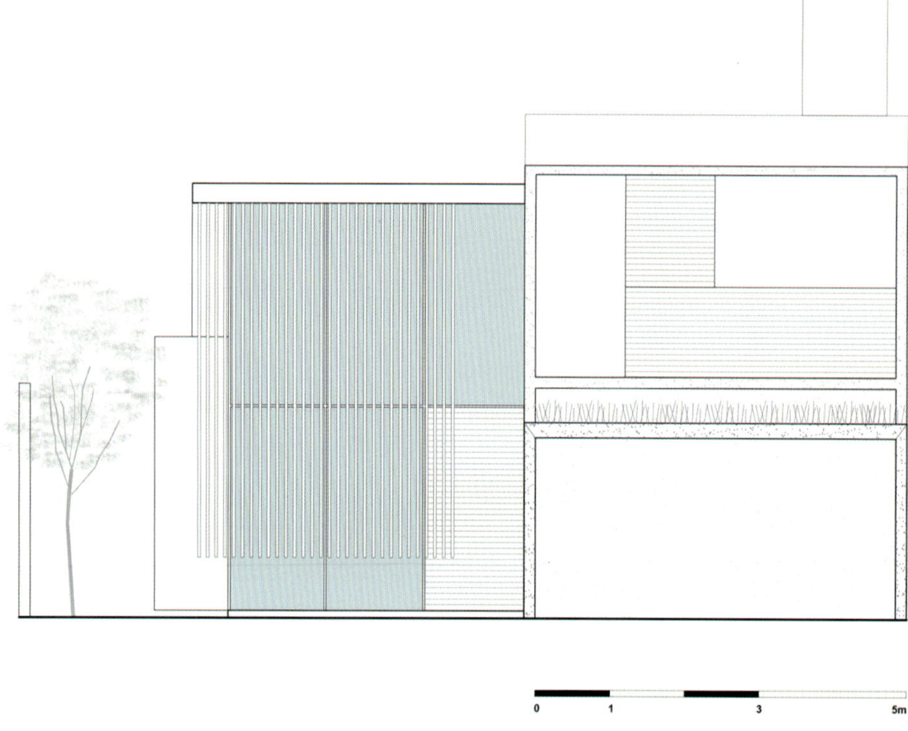

Front Elevation

CASA OSA, Costa Rica

OBRA Architects

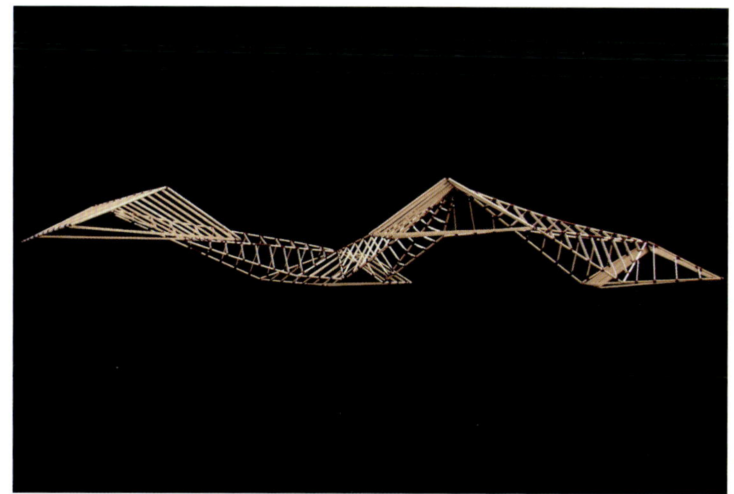

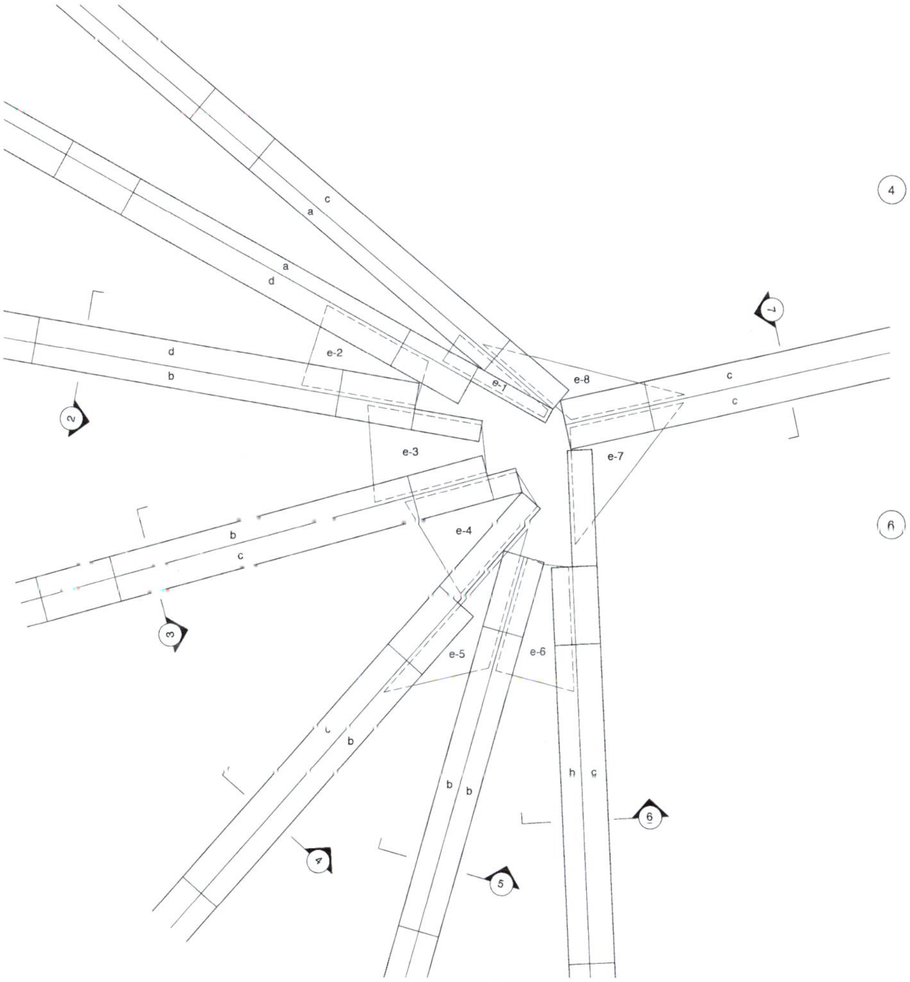

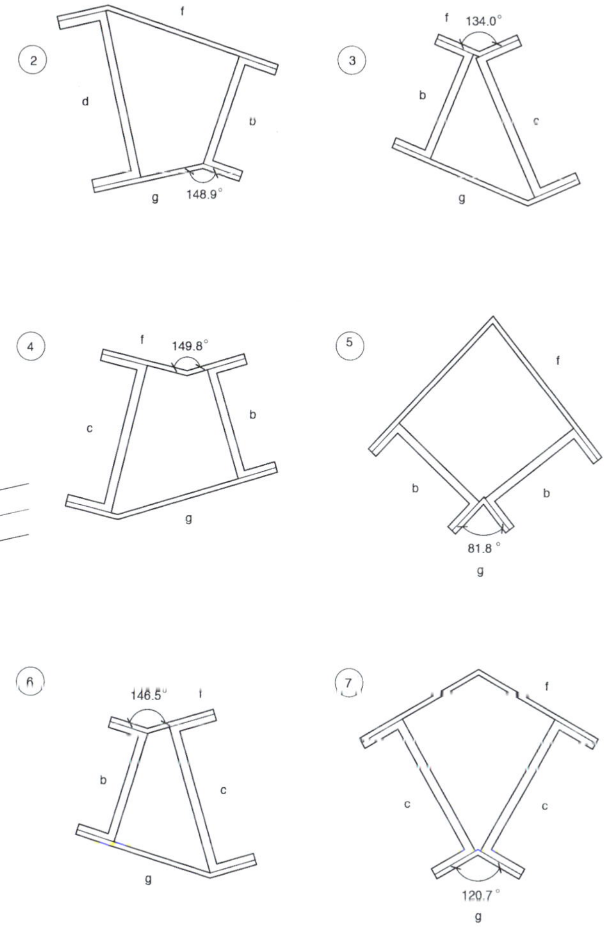

Roof Deatil

a. 76.2 mm steel channel
b. 152 mm steel channel
c. 203 mm steel channel
d. 229 mm Steel channel
e (1-8). 12.7 mm steel bracing
f. 6.5 mm angled steel plate, top
g. 6.5 mm angled steel plate, bottom

CASA PEDROSO, Mar Azul, Argentina

BAK arquitectos architecture studio

© Gustavo Sosa Pinilla

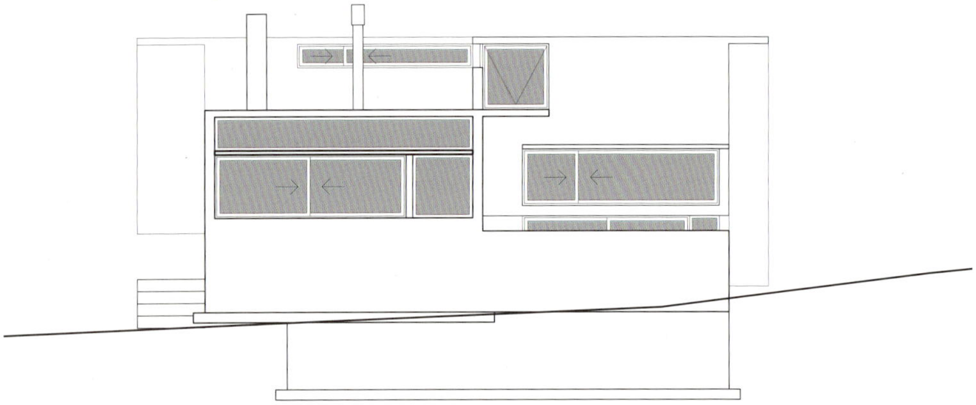

Elevation A

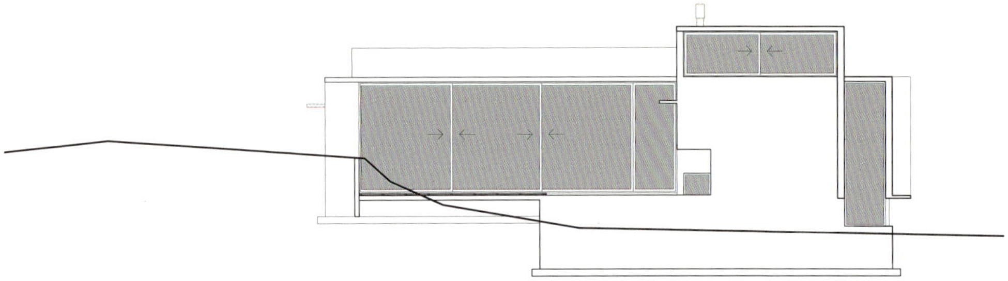

Elevation B

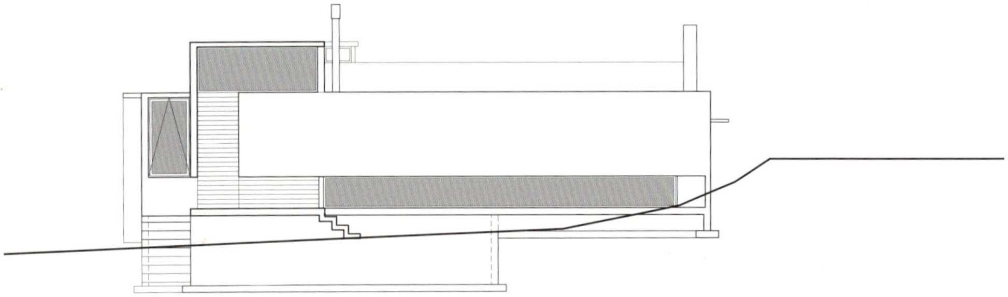

Elevation C

© Gustavo Sosa Pinilla

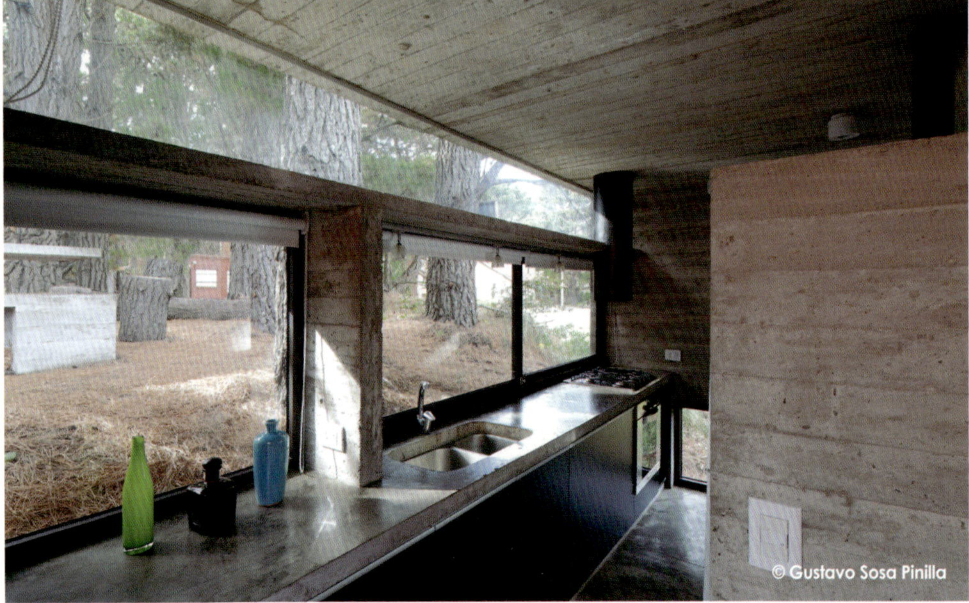

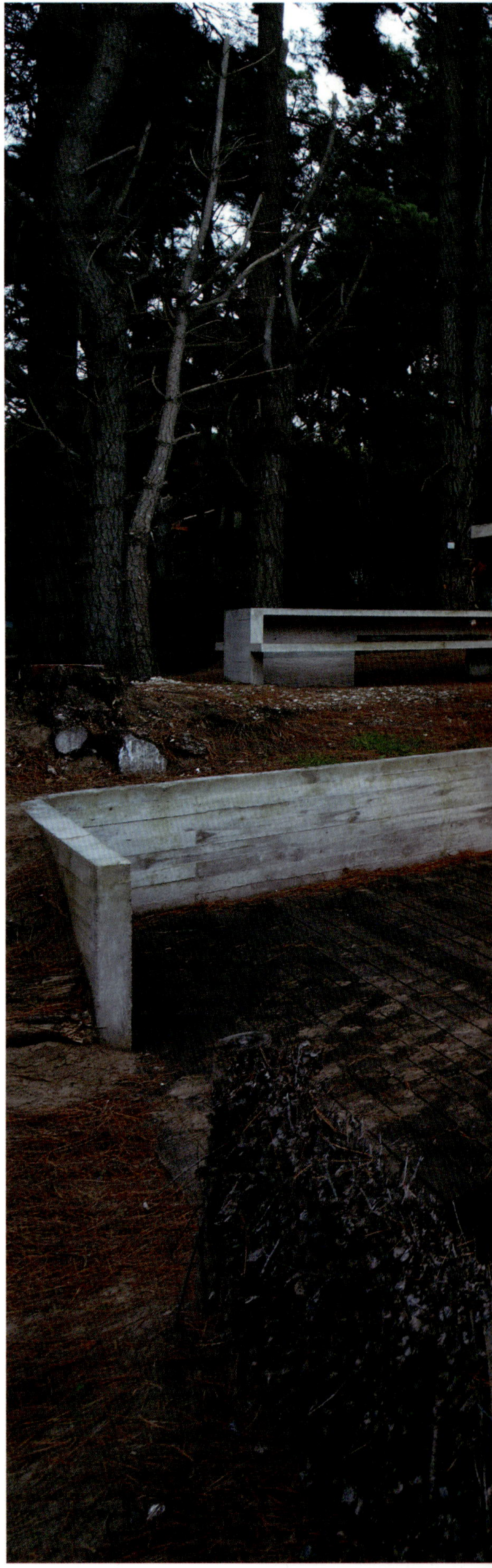

© Gustavo Sosa Pinilla

GP Mountain House, Hohenthurn, Austria

GEZA

© Massimo Crivellari

© Massimo Crivellari

© Massimo Crivellari

© Massimo Crivellari

Hankai House, Hyogo, Japan

Katsuhiro Miyamoto

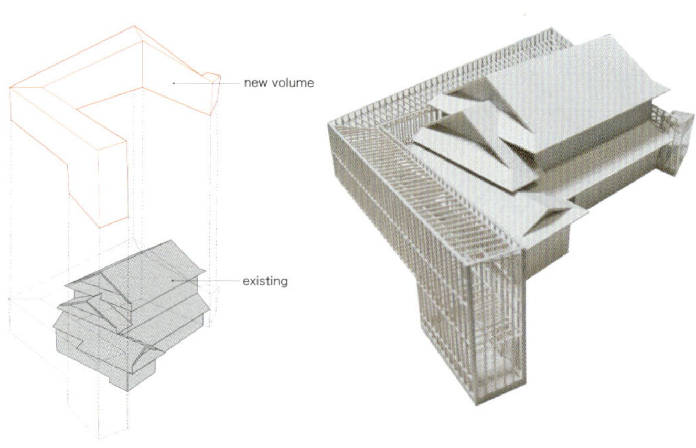

site S=1:300

Renovation of an Old House

Renovation of an Old House

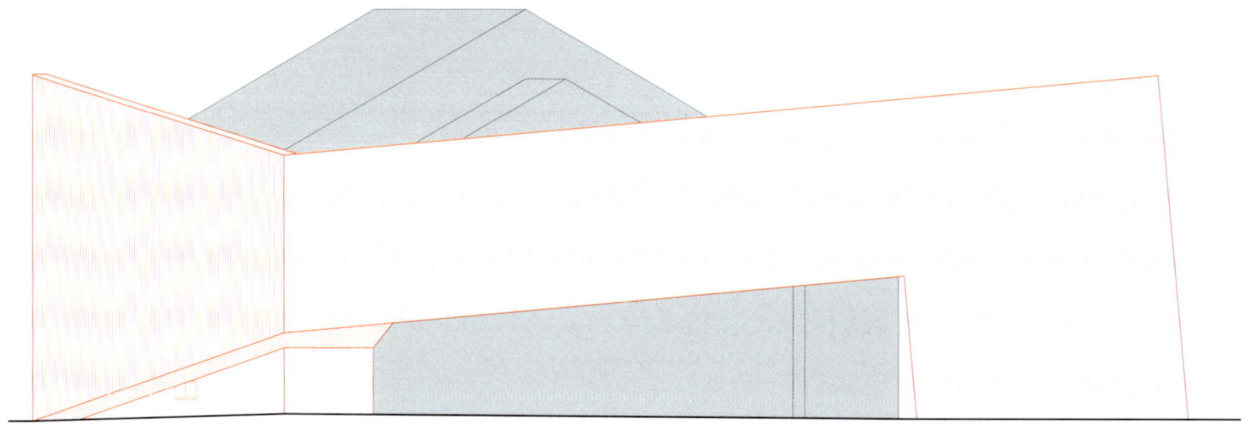

West Elevation

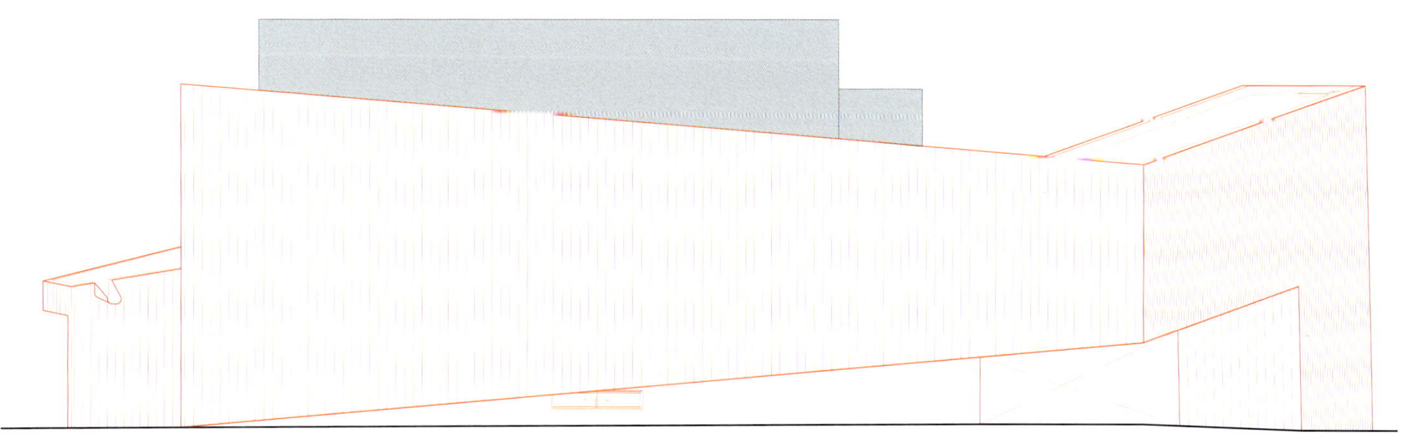

North Elevation

Sorth Elevation

Construction Process

HAUS+, Hameln, Germany

Anne Menke

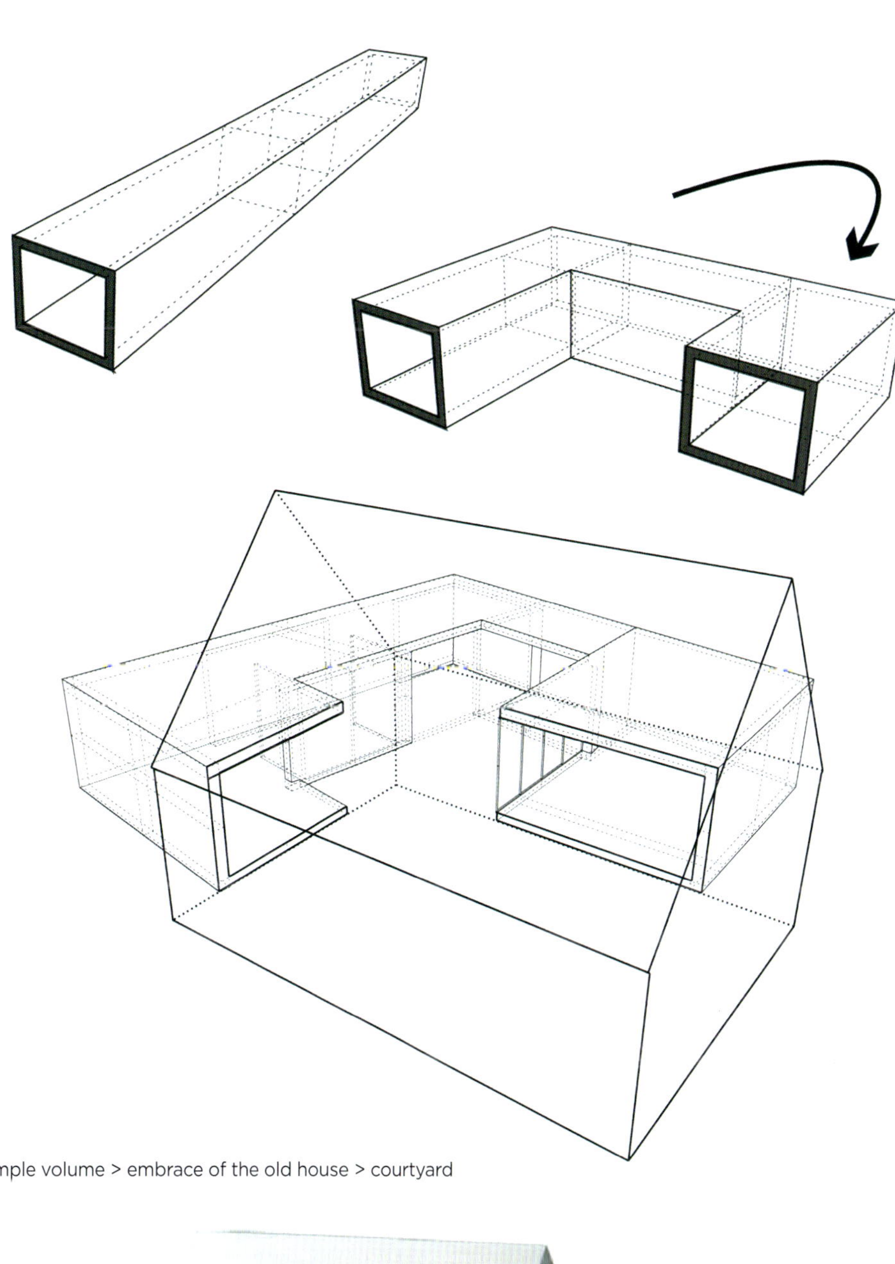

simple volume > embrace of the old house > courtyard

Study Modeling

Elevation

© Monika Marasz

© Anne Menke

© Aloys Kiefer

House Bizan Shuichiro, Tokushima, Japan

Shuichiro Yoshida Architects

© Akira Yonezu

© Akira Yonezu

© Akira Yonezu

© Akira Yonezu

House in Geumsan, Geumsan, South Korea

studio GAON

© Youngchae Park

Section 2,3,4(Above) / Elevation 3,4

Elevation 1, 2(Above) / Section 1

© Youngchae Park

© Youngchae Park

© Youngchae Park

© Youngchae Park

© Youngchae Park

© Youngchae Park

© Youngchae Park

© Youngchae Park

© Youngchae Park

© Youngchae Park

© Youngchae Park

© Youngchae Park

© Youngchae Park

© Youngchae Park

© Youngchae Park

© Youngchae Park

© Youngchae Park

© Youngchae Park

© Youngchae Park

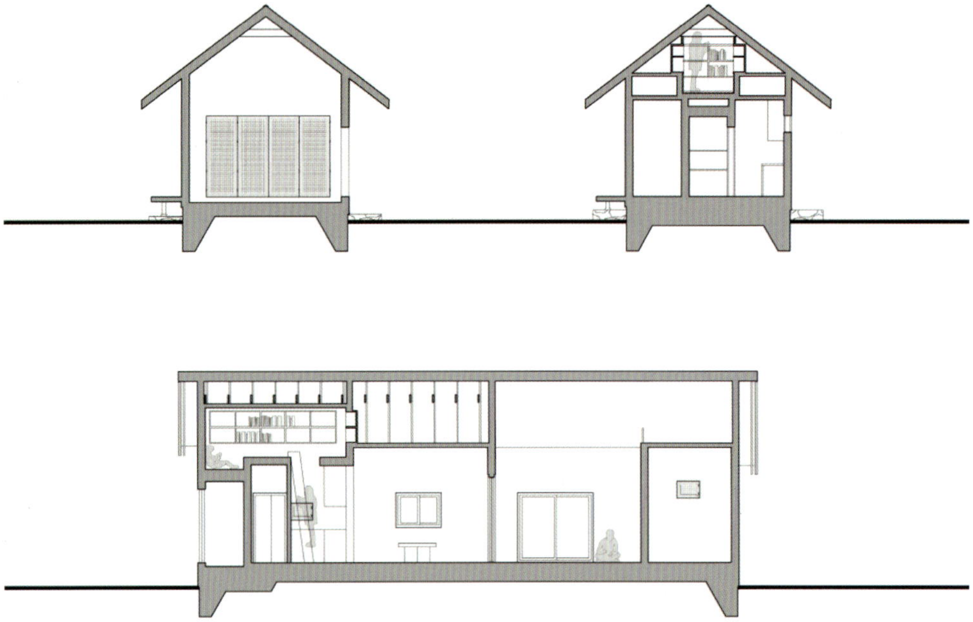

Section

Elevation

© Youngchae Park

House in Yeoju House for Reunion, Yeoju, South Korea

studio GAON

© Youngchae Park

© Youngchae Park

© Youngchae Park

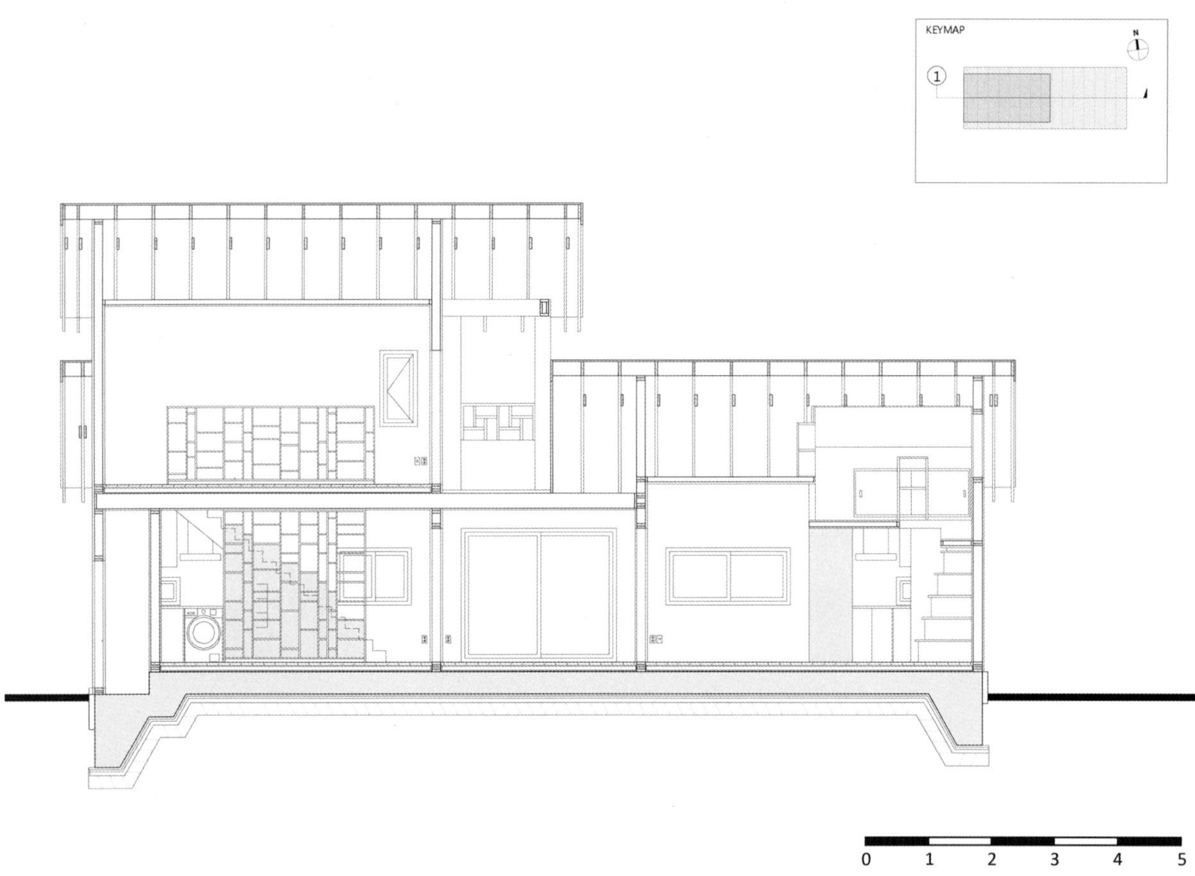

Cross Section 1

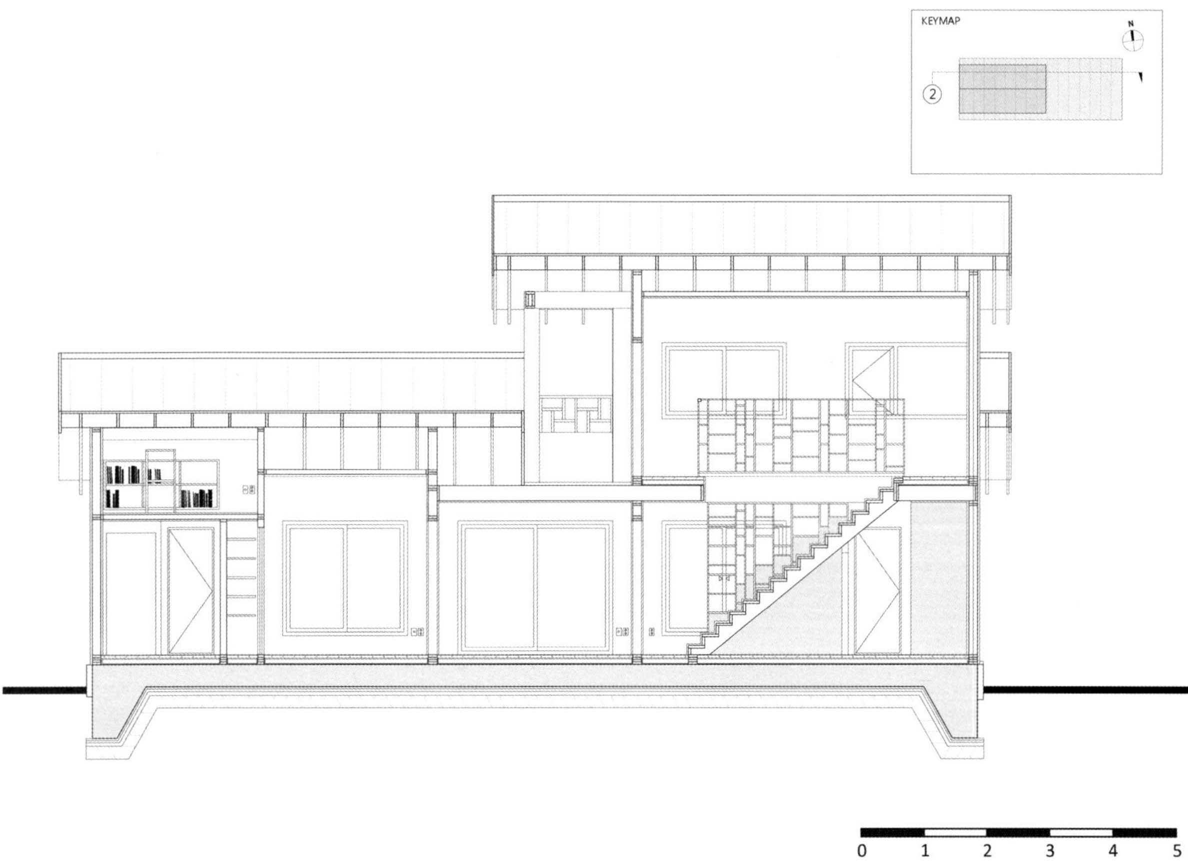

Cross Section 2

© Youngchae Park

© Youngchae Park

© Youngchae Park

South Elevation 1

South Elevation 2

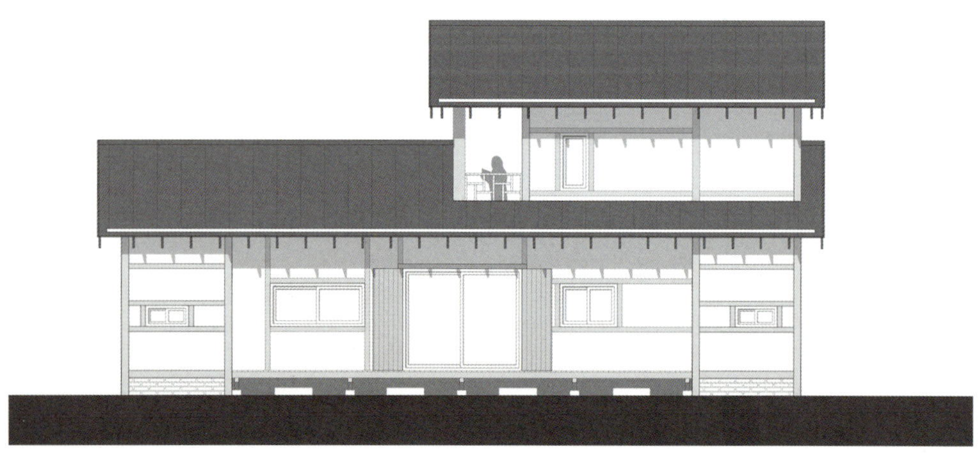
North Elevation

© Youngchae Park

House with Chapel, Innsbruck, Austria

Martin Mutschlechner, Barbara Lanz

Sketch

Study Modeling

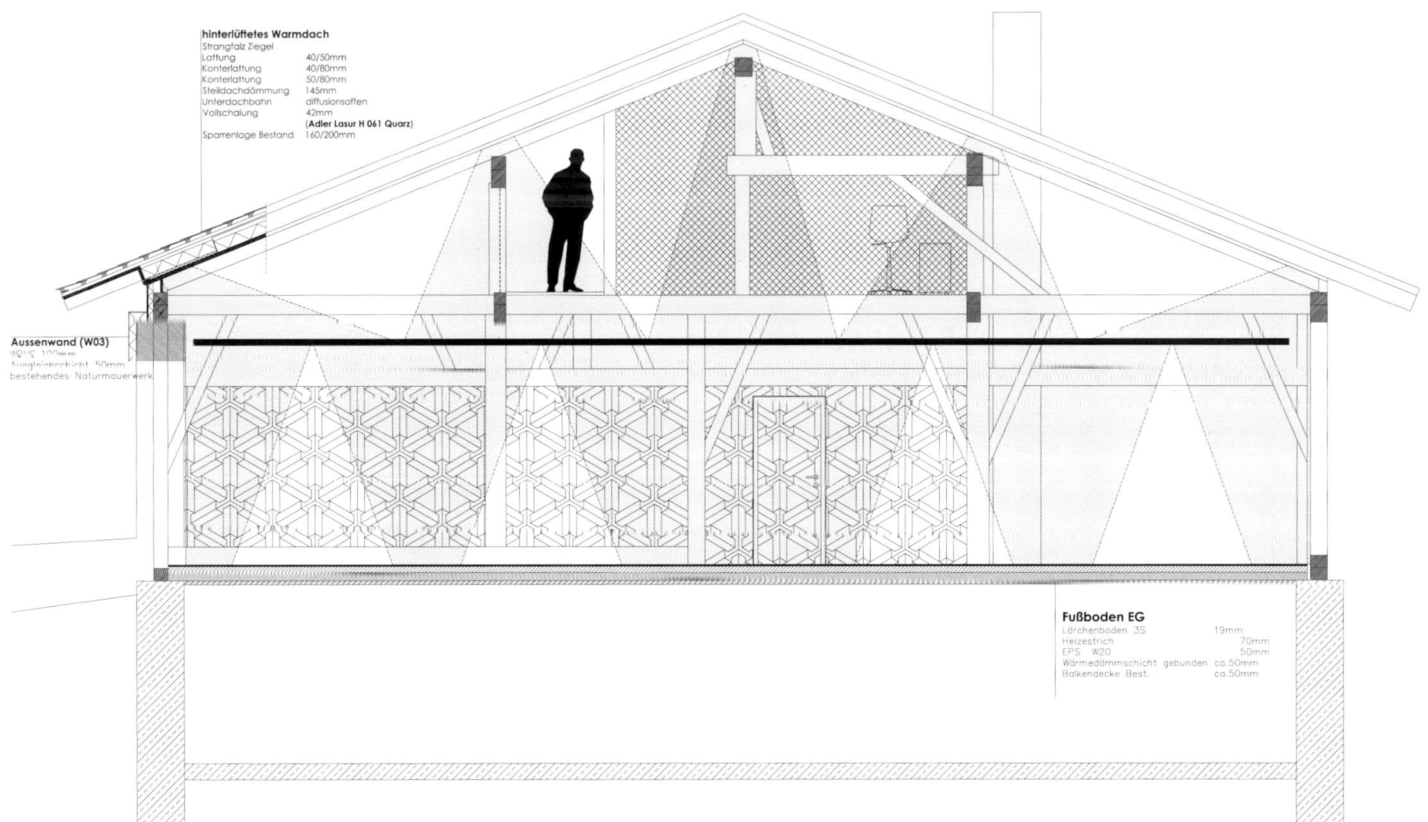

Section C1

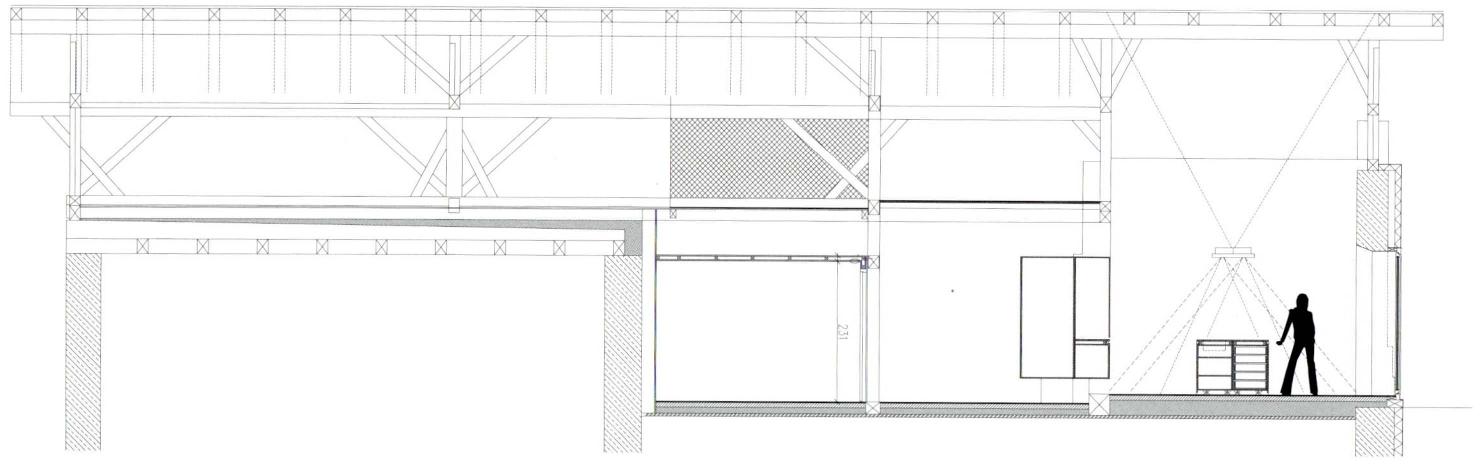

Longitudinal Section

Hunsett Mill, Norfolk, UK

Friedrich Ludewig, Stefano Dal Piva

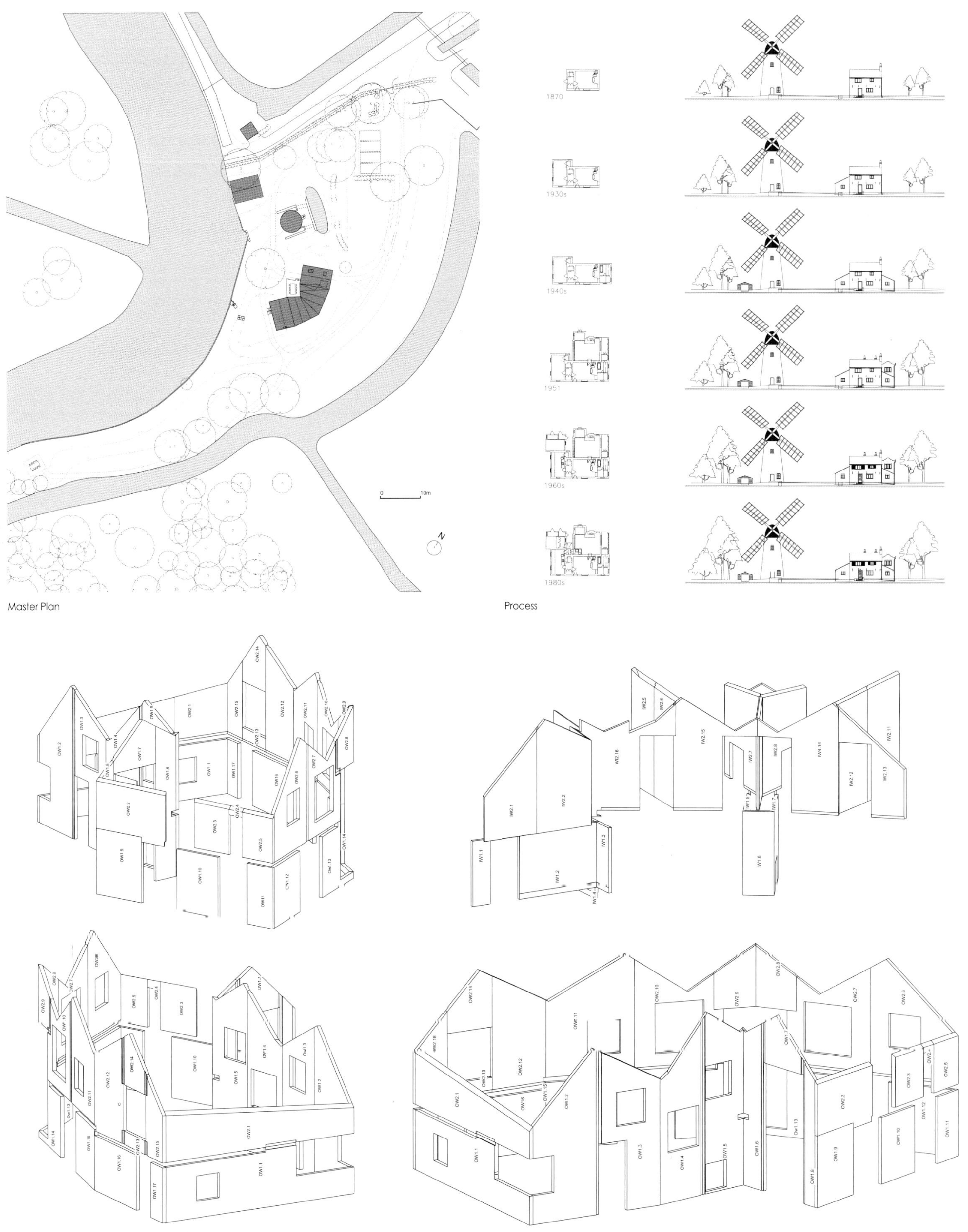

Master Plan

Process

Assembly Drawings for the Timber Structure

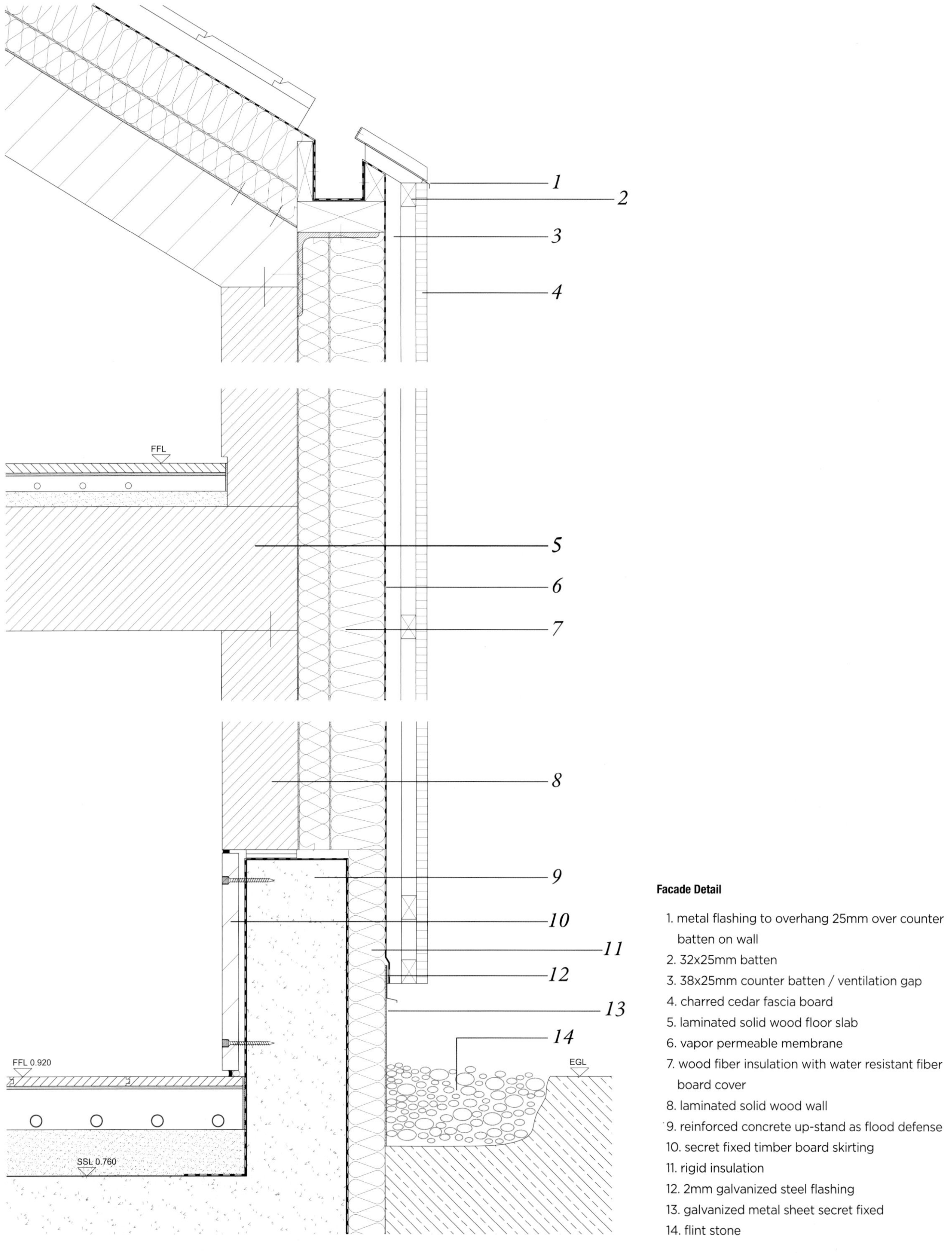

Facade Detail

1. metal flashing to overhang 25mm over counter batten on wall
2. 32x25mm batten
3. 38x25mm counter batten / ventilation gap
4. charred cedar fascia board
5. laminated solid wood floor slab
6. vapor permeable membrane
7. wood fiber insulation with water resistant fiber board cover
8. laminated solid wood wall
9. reinforced concrete up-stand as flood defense
10. secret fixed timber board skirting
11. rigid insulation
12. 2mm galvanized steel flashing
13. galvanized metal sheet secret fixed
14. flint stone

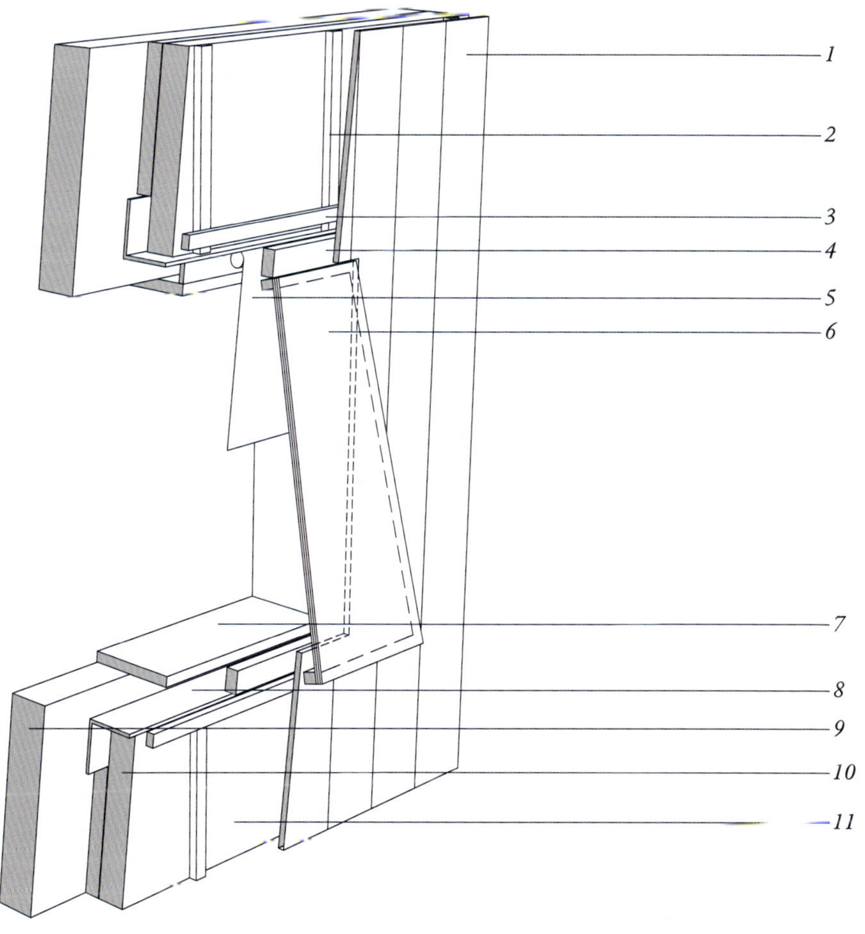

Window Detail

1. charred larch cladding
2. 38x25mm counter batten
3. 38x25mm batten
4. timber window frame
5. recessed roller blind
6. openable double glazing unit structurally bonded to frame
7. soft wood lining
8. galvanized metal frame to support glazing frame
9. laminated solid wood
10. wood fiber insulation with water resistant fiber board cover
11. vapor permeable membrane

Koro House, Toyota Aichi, Japan

Katsutoshi Sasaki + Associates

© Katsutoshi Sasaki + Associates

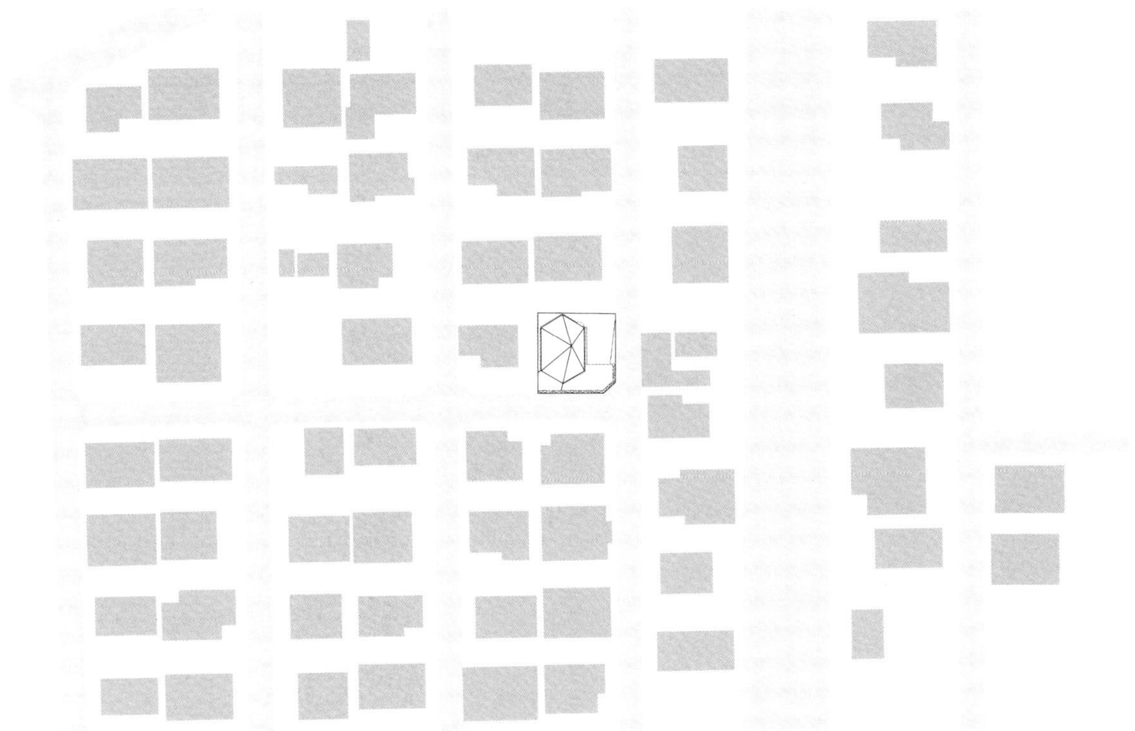

Site Plan

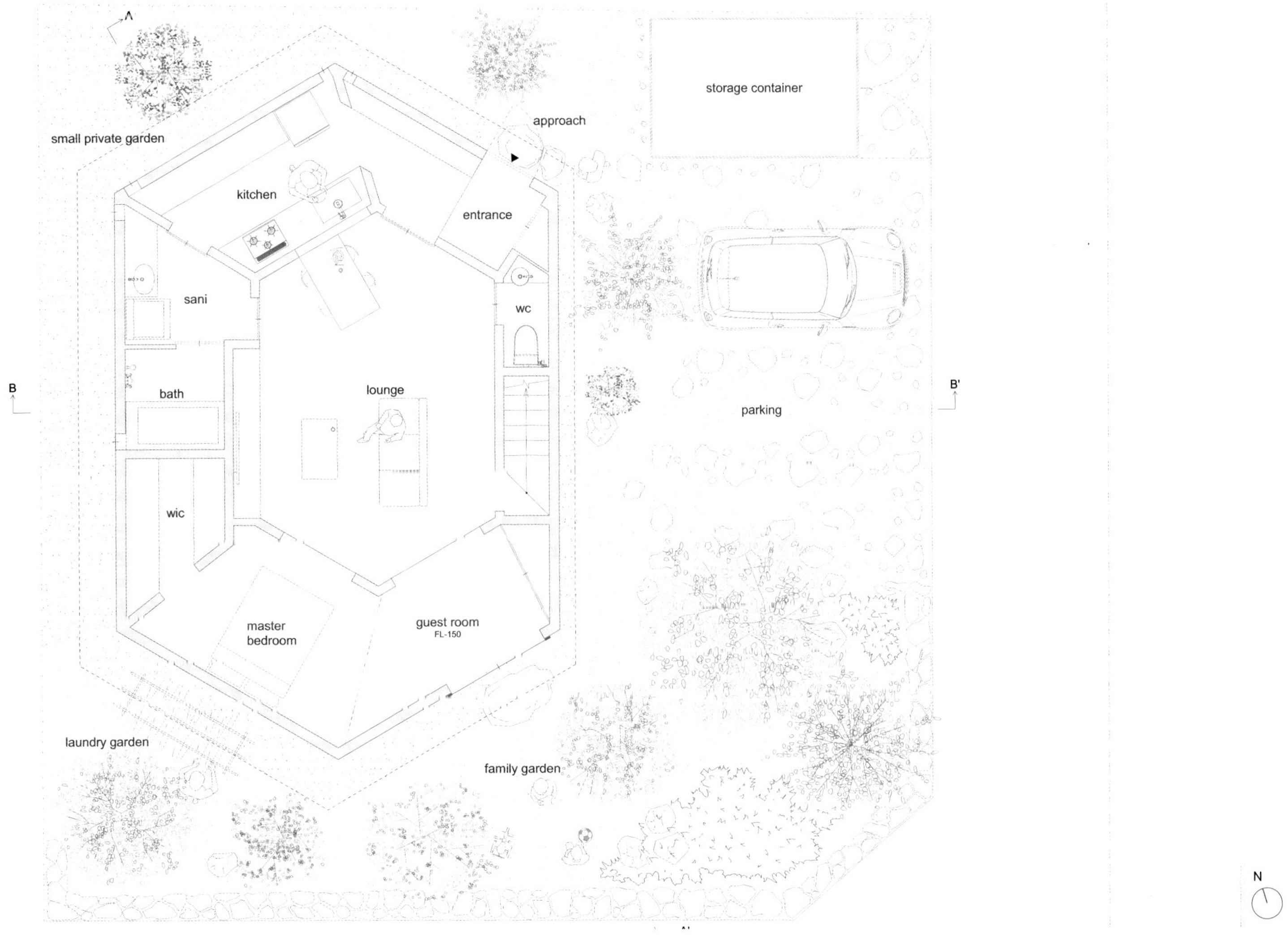

A
small private garden
approach
storage container
kitchen
entrance
sani
wc
B
bath
lounge
B'
parking
wic
master
bedroom
guest room
FL-150
laundry garden
family garden
N

Floor Plan

© Katsutoshi Sasaki + Associates

© Katsutoshi Sasaki + Associates

© Katsutoshi Sasaki + Associates

© Katsutoshi Sasaki + Associates

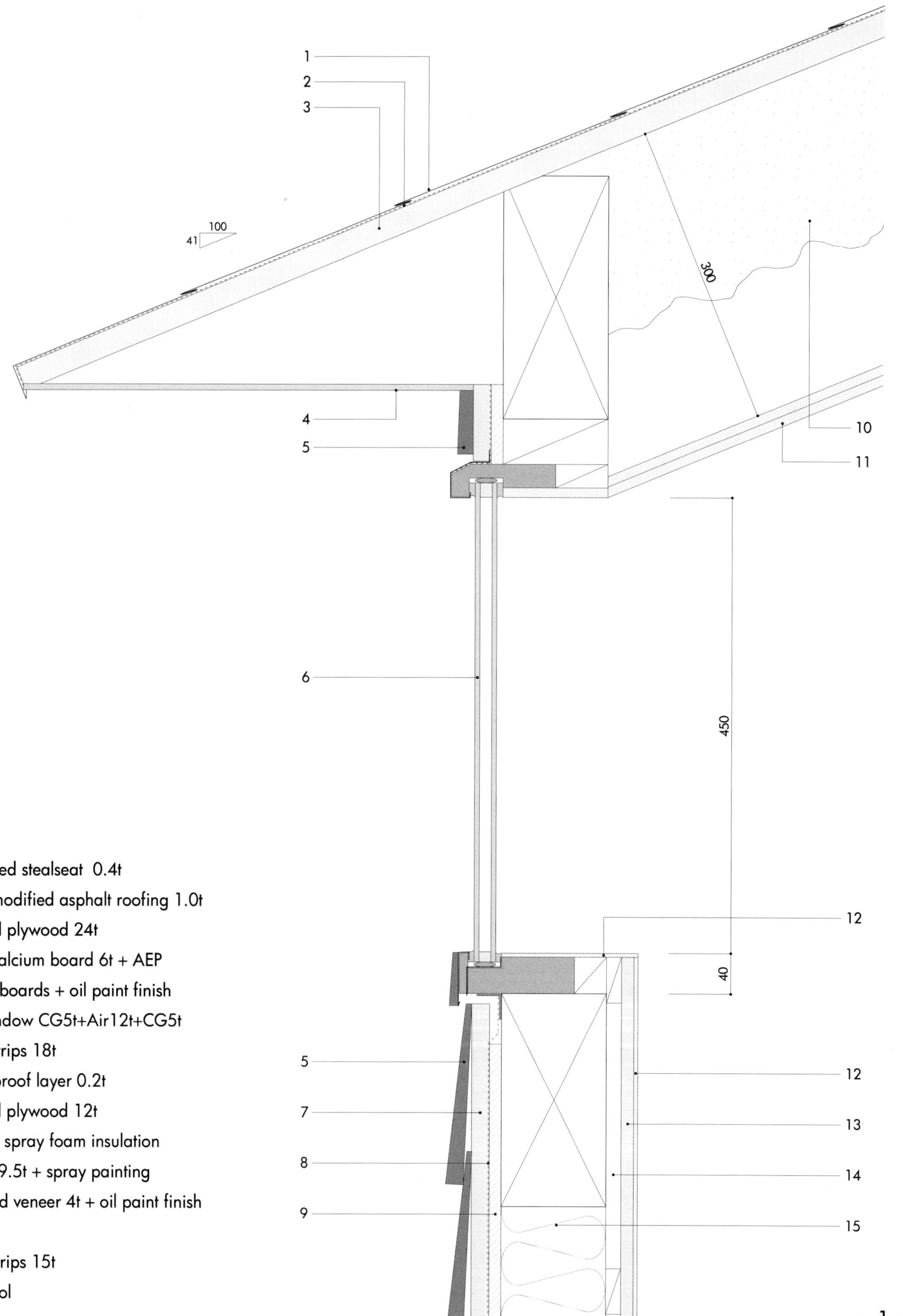

1. garvanised stealseat 0.4t
2. rubber-modified asphalt roofing 1.0t
3. structural plywood 24t
4. silicate calcium board 6t + AEP
5. wooden boards + oil paint finish
6. fixed window CG5t+Air12t+CG5t
7. furring strips 18t
8. vapour proof layer 0.2t
9. structural plywood 12t
10. urethane spray foam insulation
11. PB 9.5t+9.5t + spray painting
12. decorated veneer 4t + oil paint finish
13. PB 12.5t
14. furring strips 15t
15. glass wool

Highside Detail

Construction Process

© Katsutoshi Sasaki + Associates

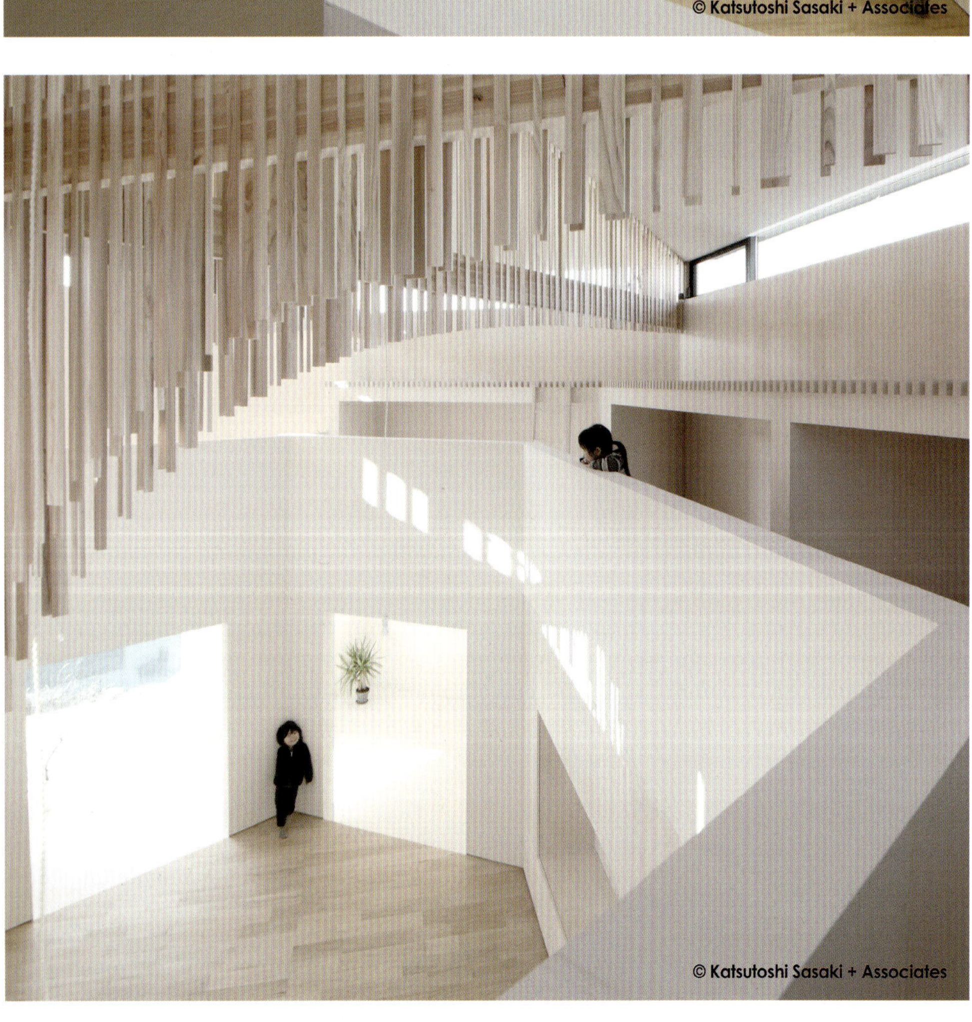
© Katsutoshi Sasaki + Associates

© Katsutoshi Sasaki + Associates

Maison Leguay, Palencia, Spain

Moussafir Architectes Associés

© Jérôme Ricolleau

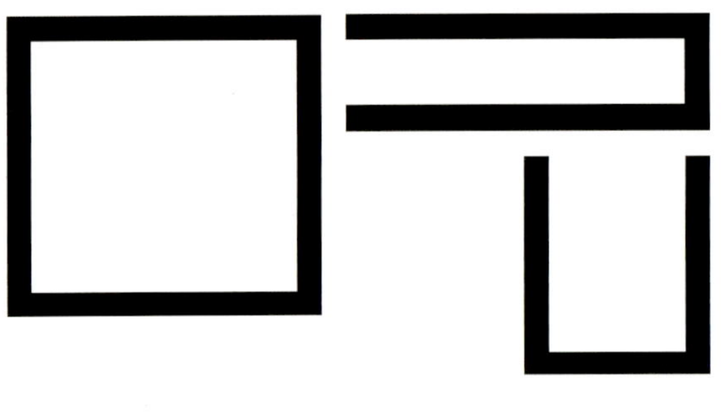

Cloning a House

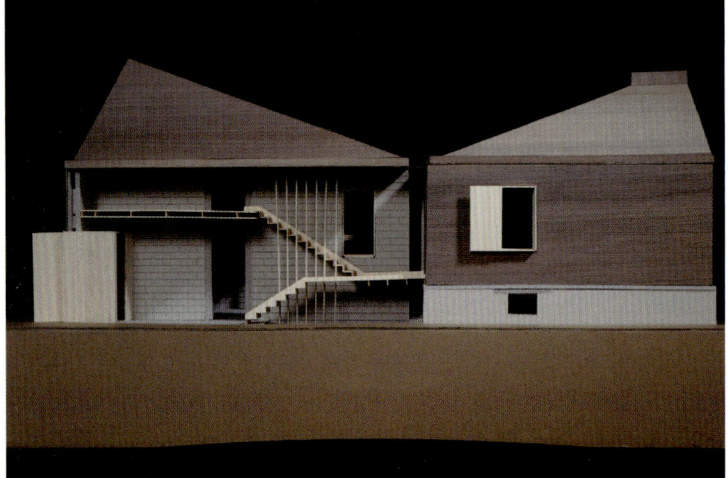

Study Modeling

Construction Process

3.

4.

5.

6.

© Jérôme Ricolleau

© Jérôme Ricolleau

© Jérôme Ricolleau

© Jérôme Ricolleau

Metamorphosis, Casablanca, Chile

Delphine Ding, José Ulloa Davet

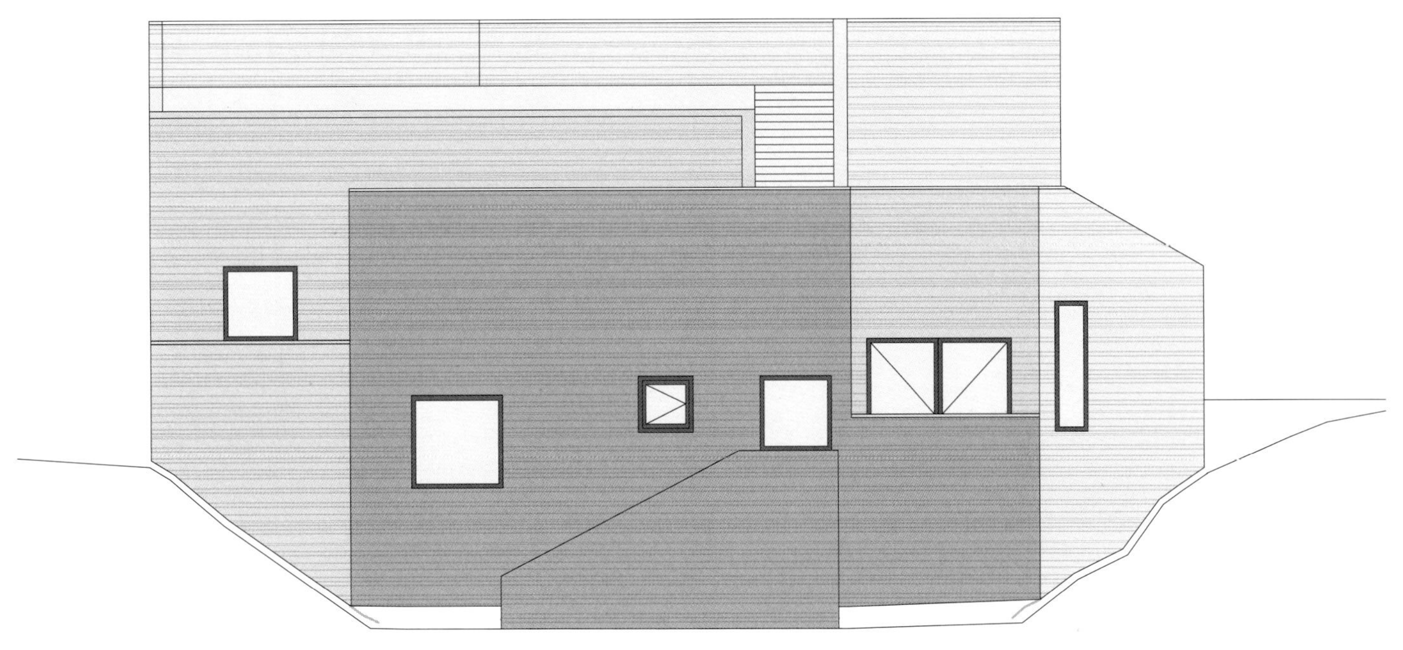

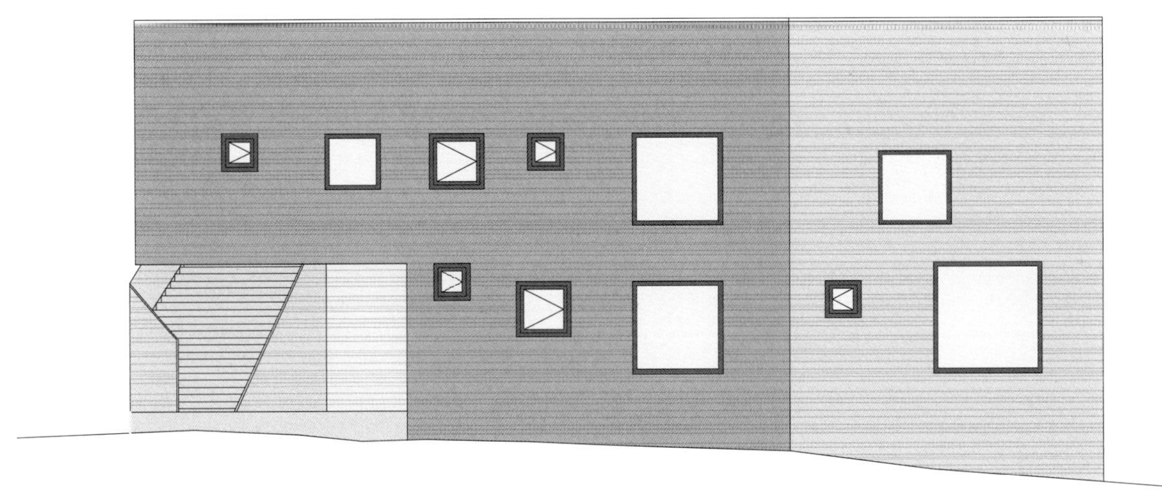

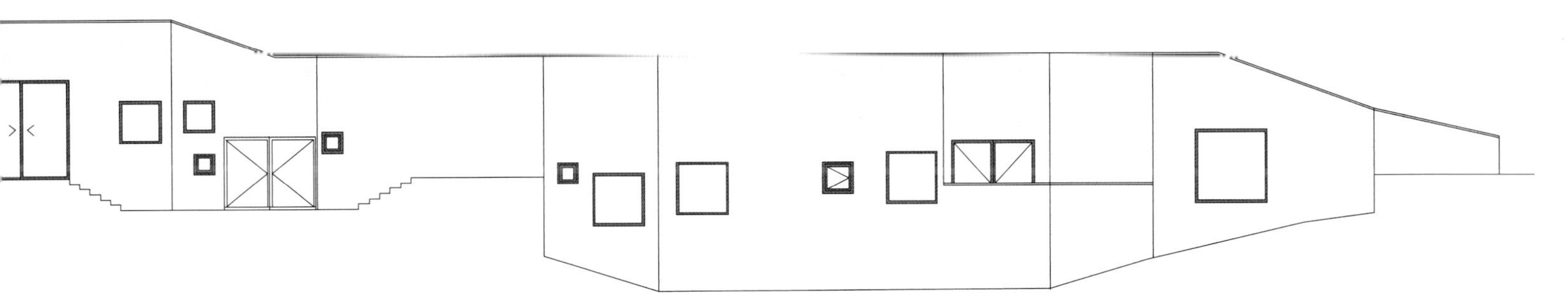

Detail

1. roof:
1" x 4" timber planks, 2" x 2" battens
bituminous sheeting
12mm laminated wood board
100mm rockwool isolation
vapor barrier
12mm laminated wood board

2. exterior walls:
1" x 4" timber planks,
2" x 2" battens / air chamber
breathable and wet proof layer
12mm laminated wood board
100mm rockwool isolation
vapor barrier
12mm laminated wood board
1" x 4" timber planks

3.
1" x 4" assembled timber planks
12mm laminated wood board
100mm rockwool isolation
12mm laminated wood board
10" double Oregon pine beams for cantilever

4.
1" x 4" assembled timber planks
vapor barrier
12mm laminated wood board
100mm rockwool isolation
12mm laminated wood board

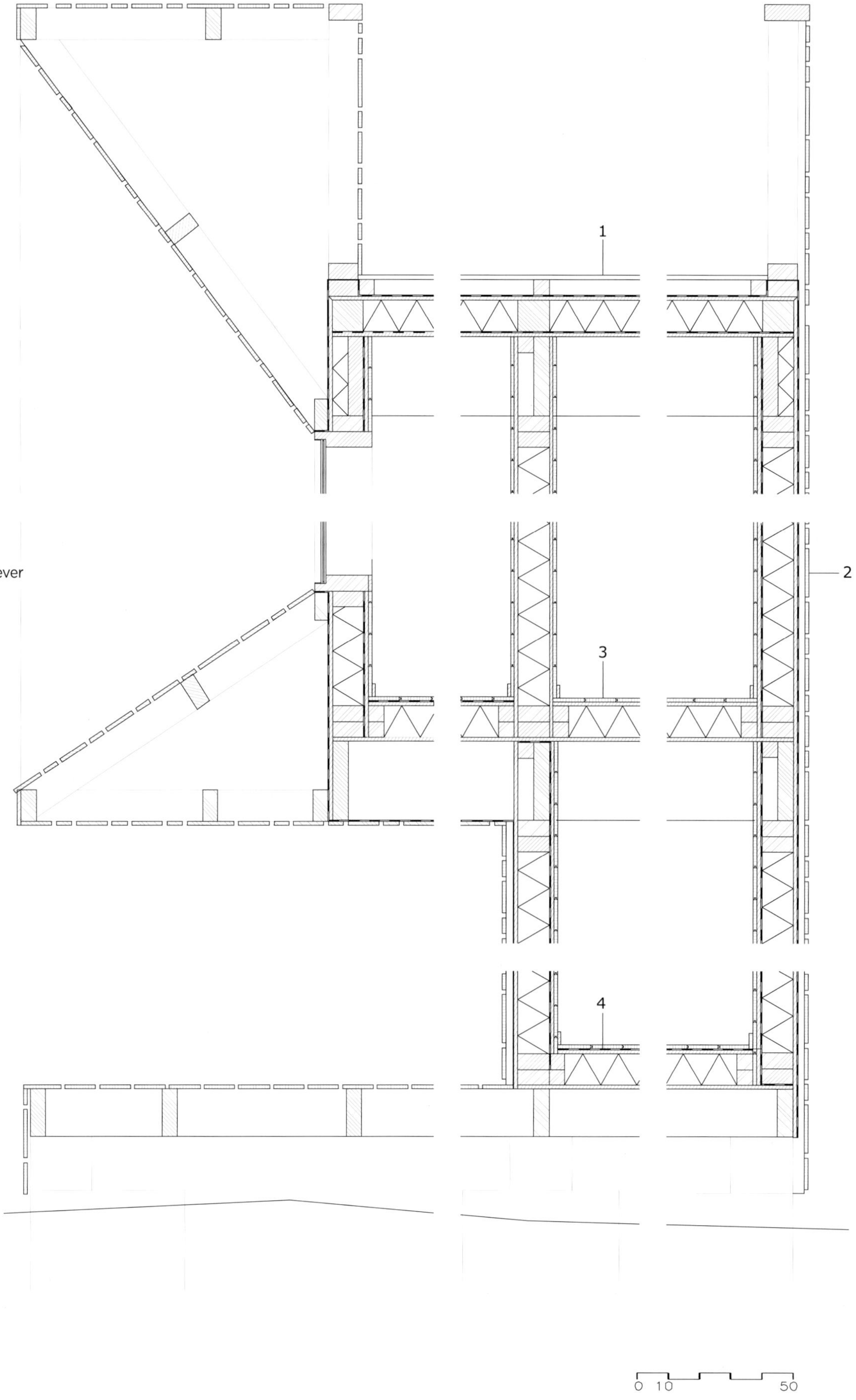

Before

1.

2.

3.

4.

5.

Construction Process

Muslhaufen House, Luson, Italy

Armin Blasbichler Studio

Study Modeling

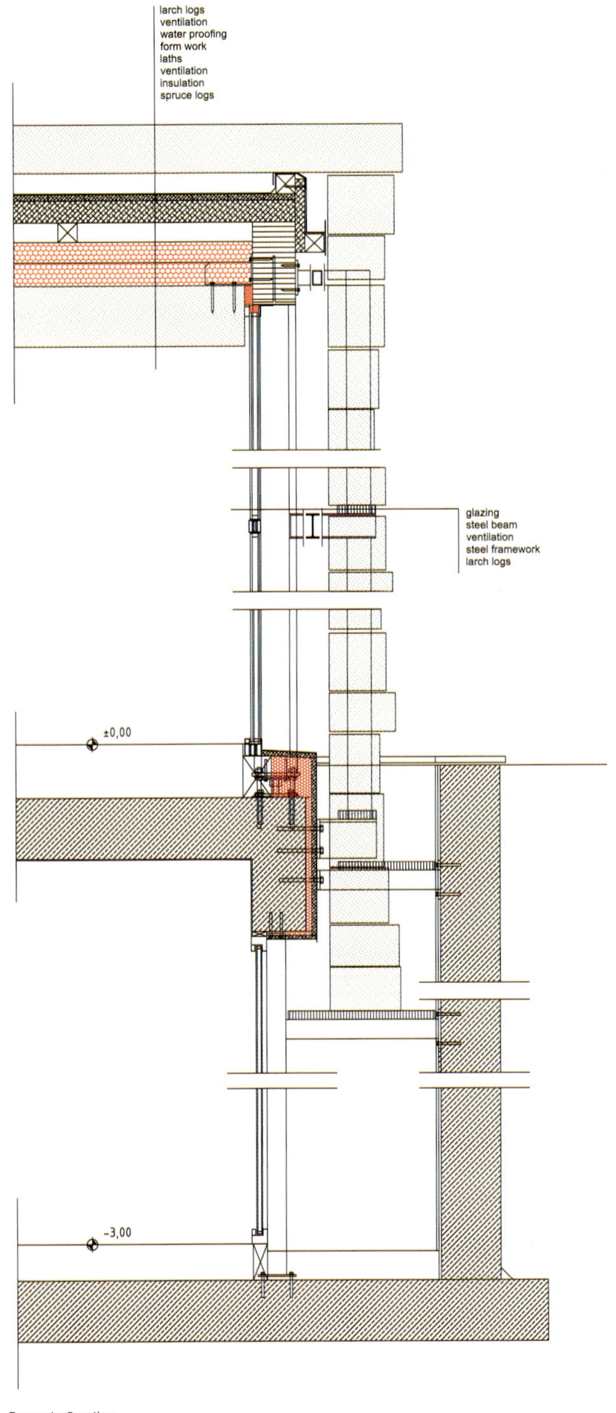

Facade Section

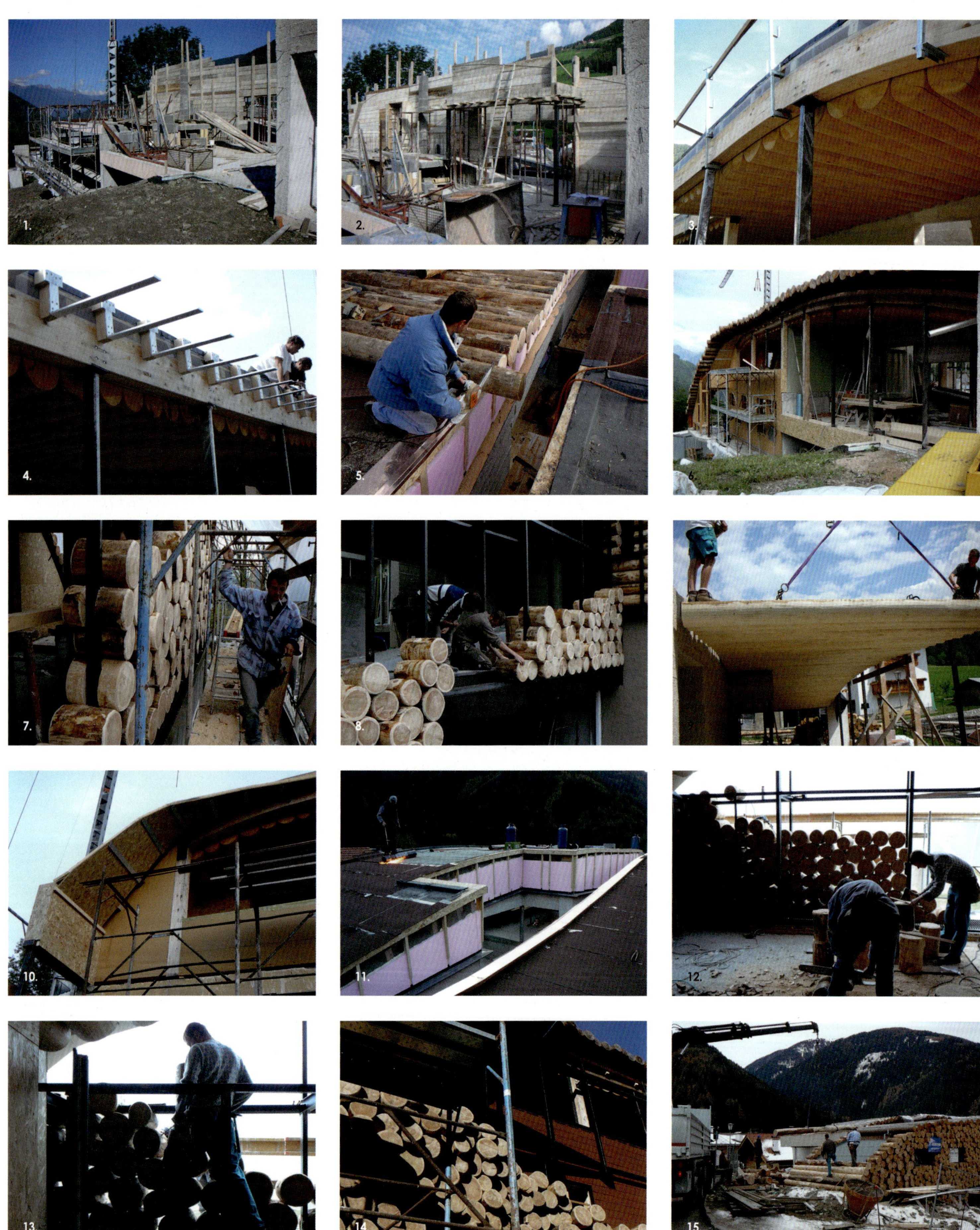

Construction Process

© Ingrid Heiss

© Ingrid Heiss

© Ingrid Heiss

Mylla Hytte, Jevnaker County, Norway
Mork-Ulnes Architects

© Bruce Damonte

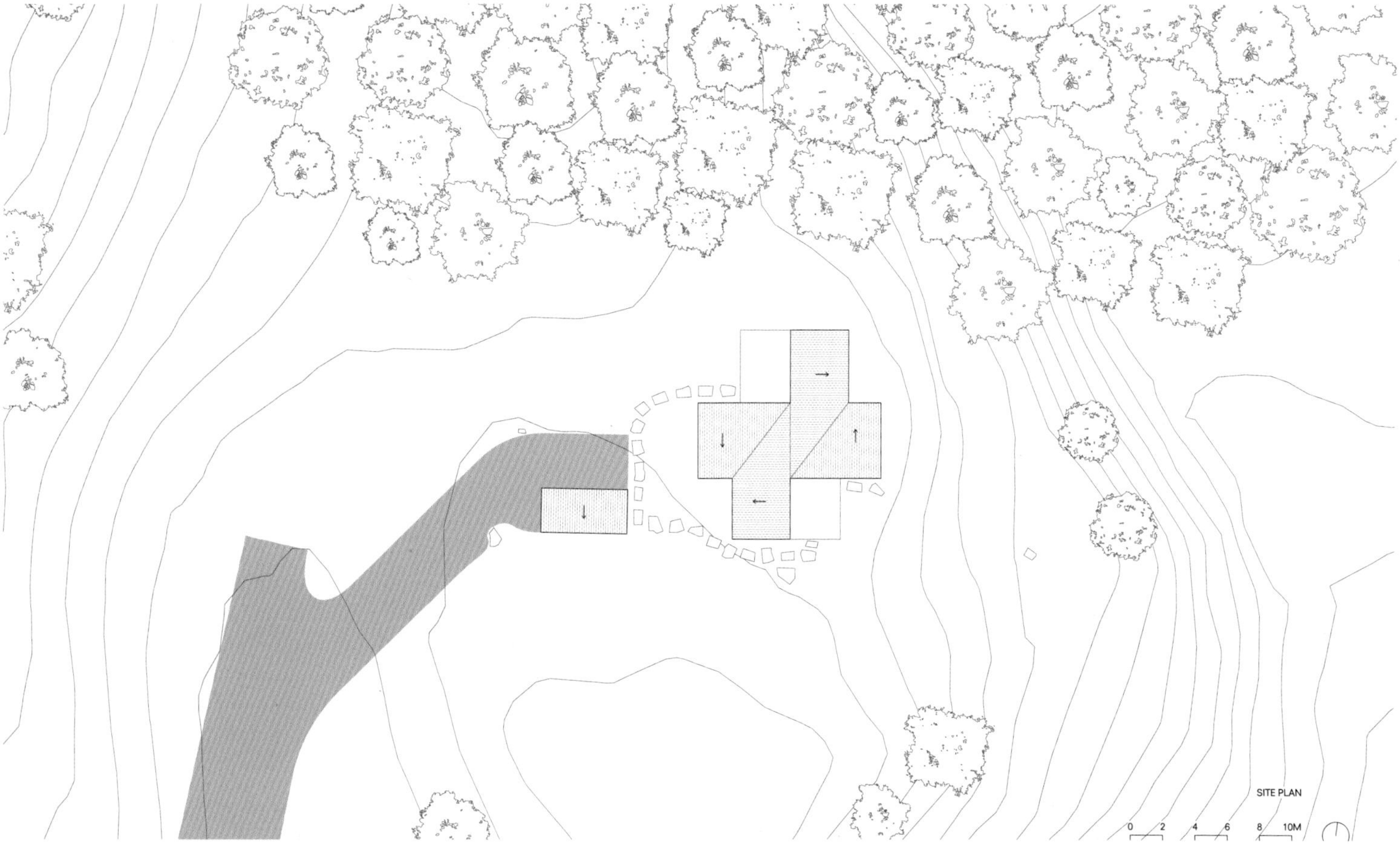

SITE PLAN

0 2 4 6 8 10M

Site Plan

Section A

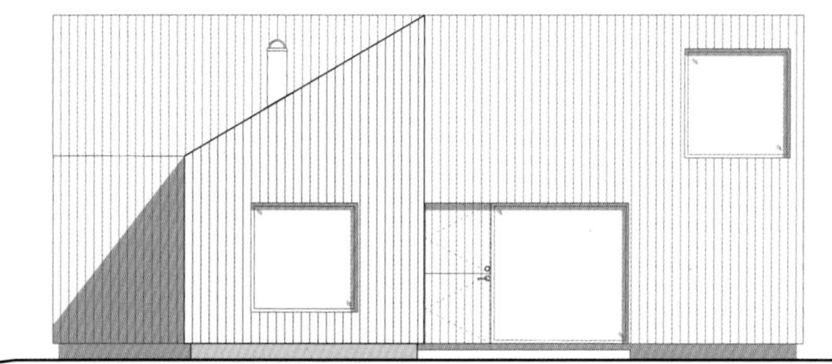
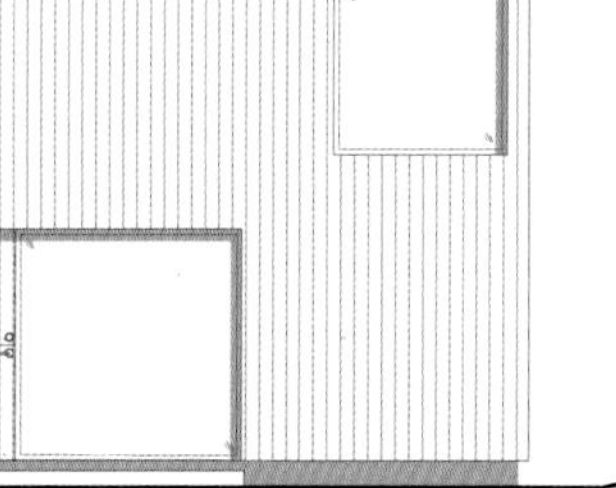

North Elevation

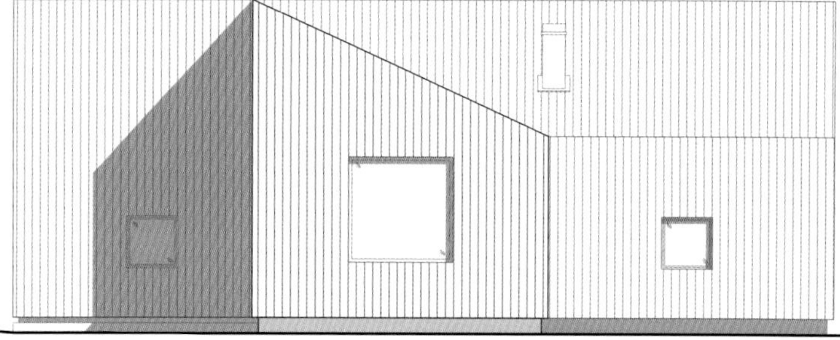

East Elevation

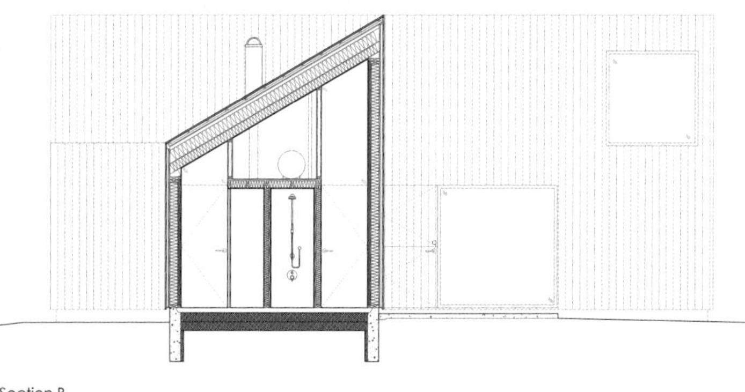

Section B

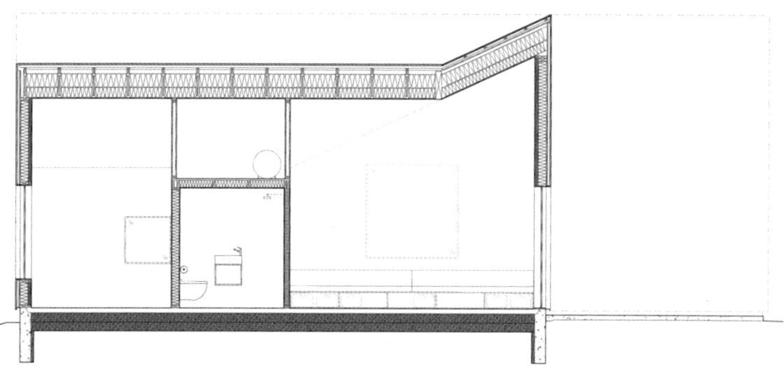

Section C

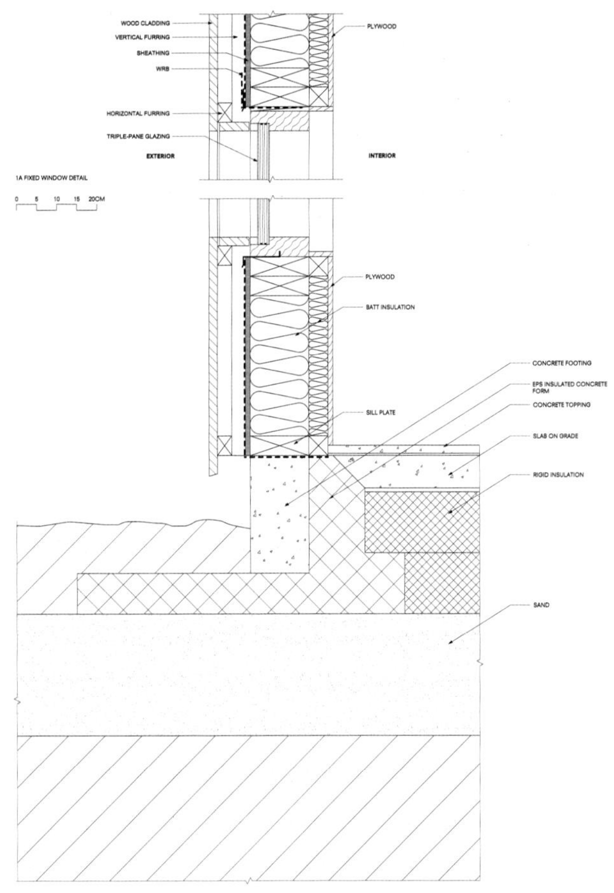

1B Fixed Window Detail with Foundation

0 5 10 15 20CM

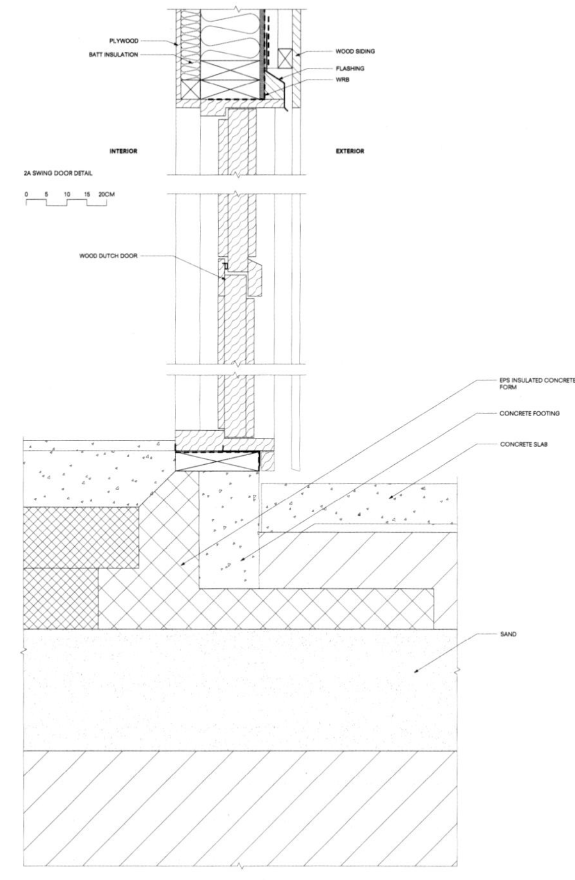

2B Swing Door Detail with Foundation

0 5 10 15 20CM

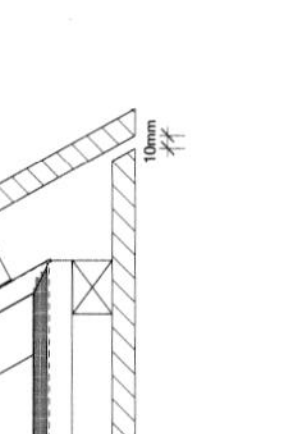

21X148MM KEBONY TAKBORD
36MM LEKTER
23MM SLØYFER
UNDERTAKSPAPP / UV DUK
19MM RUPANEL
48MM PÅFORING
VINDSPERRE / TYVEK
350MM I-PROFIL - ISOLERT
DAMPSPERRE
9MM FINER

21X148MM KEBONY KLEDNINGSBORD - STÅENDE
36MM LEKTER
23MM SLØYFER
UV DUK
12MM ASFALTPLATER
148MM STENDERVERK - ISOLERT
DAMPSPERRE
48MM UTLEKTING - ISOLERT
12 MM FINER

0 5 10 15 20CM

Roof Rake Detail

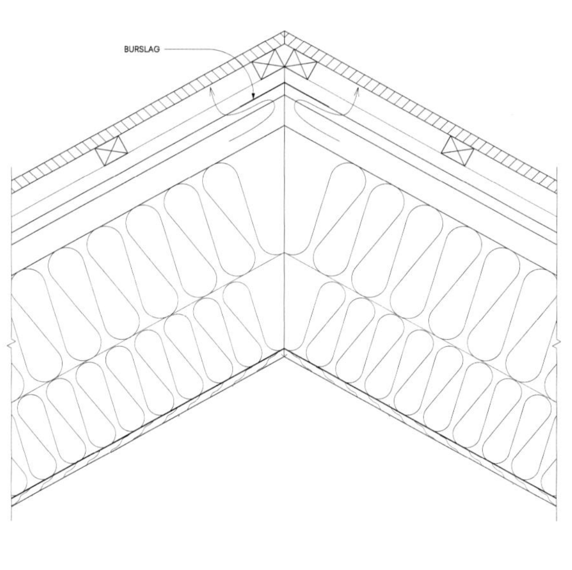

0 5 10 15 20CM

Roof Ridge Detail

© Bruce Damonte

© Bruce Damonte

© Bruce Damonte

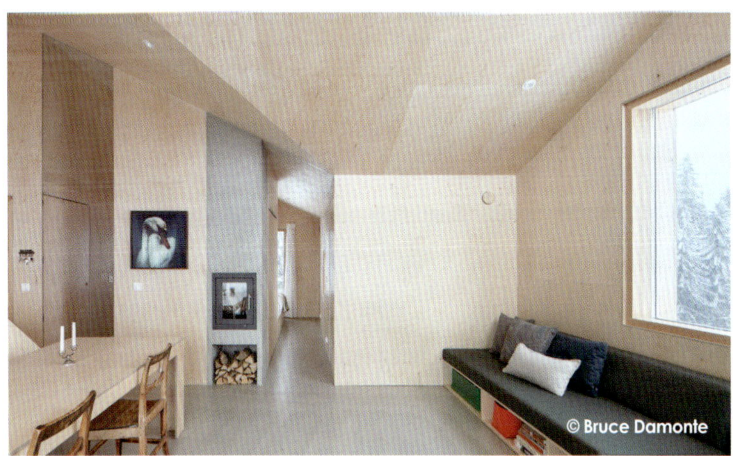

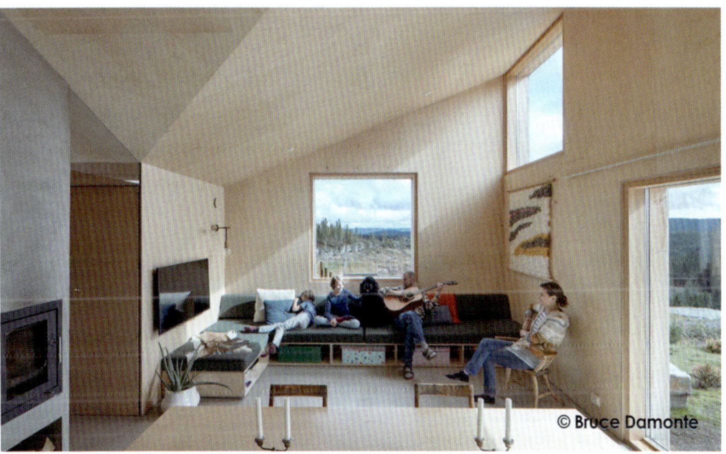

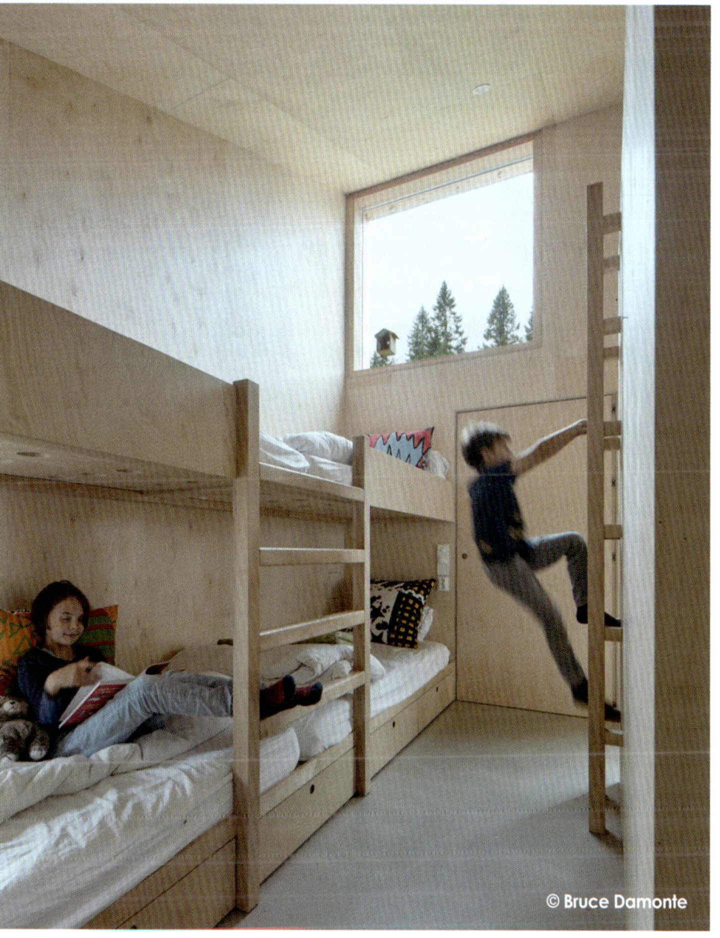

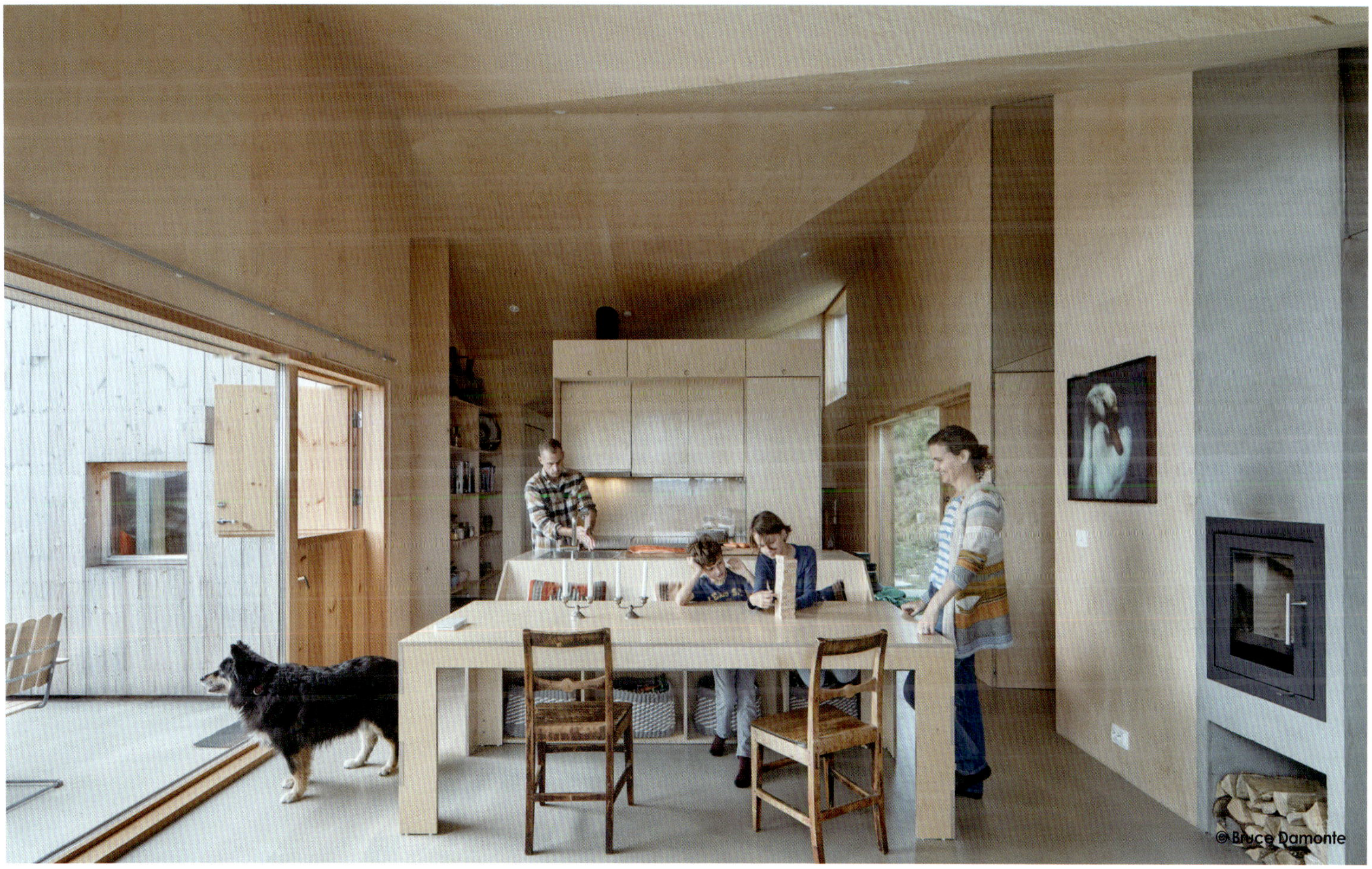
© Bruce Damonte

© Bruce Damonte

Swisshouse Rossa, Graubünden, Switzerland

Davide Macullo Architects

Idea Sketch

Site Plan

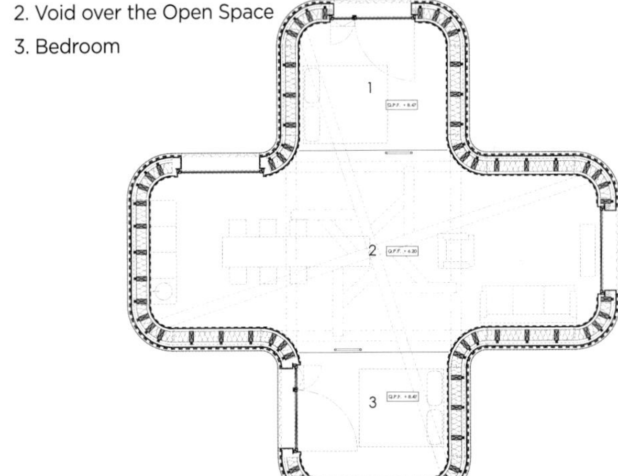

First Floor Mezzanine

Roof Plan

© Alexandre Zweiger

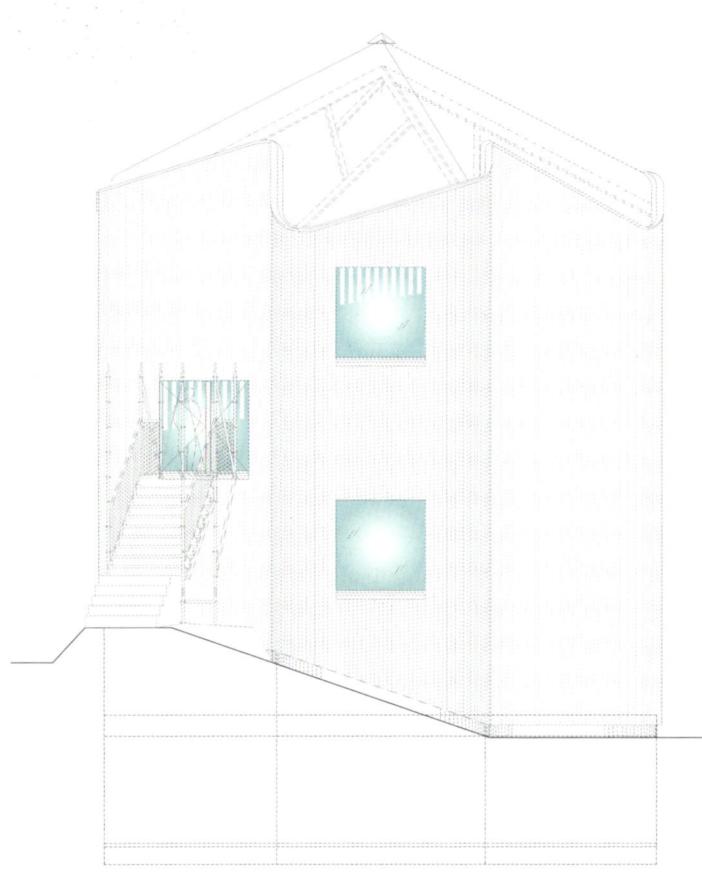
East Elevation

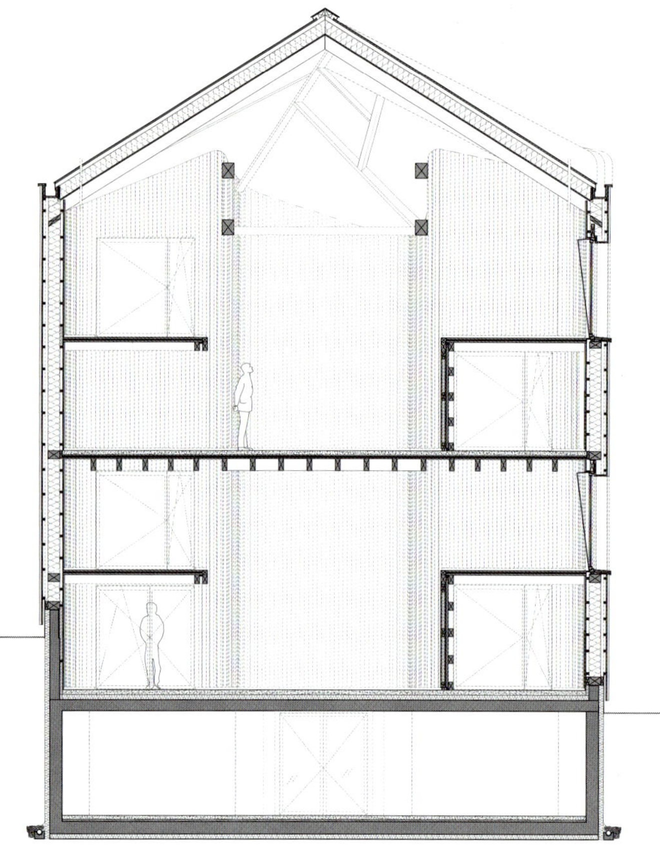
Section A

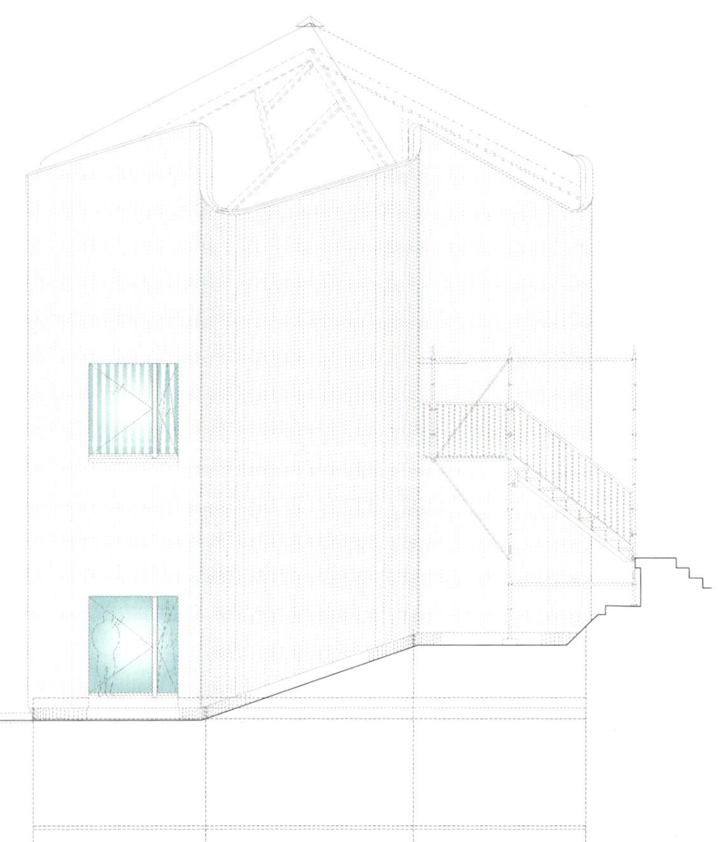
South Elevation

West Elevation

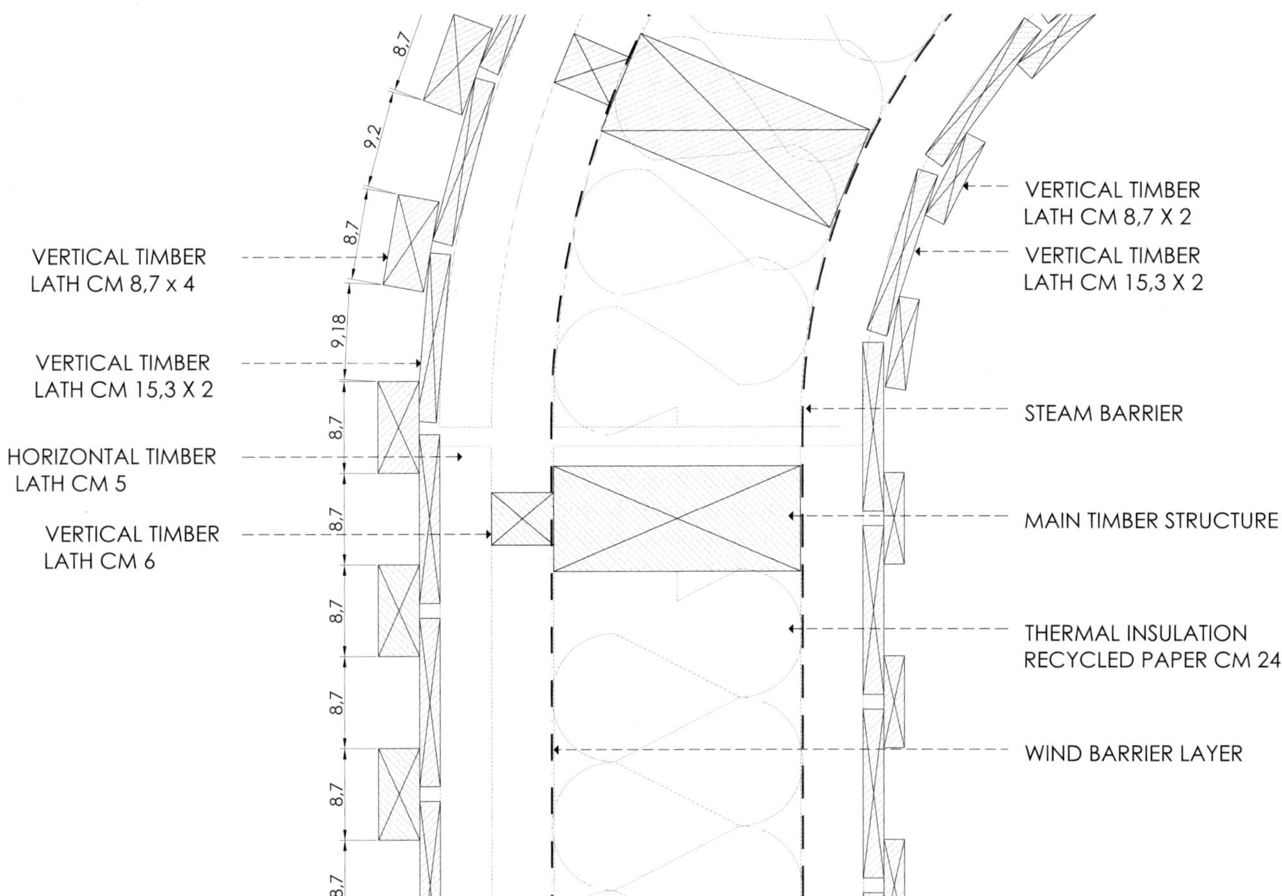

Detail - Vertical Slats

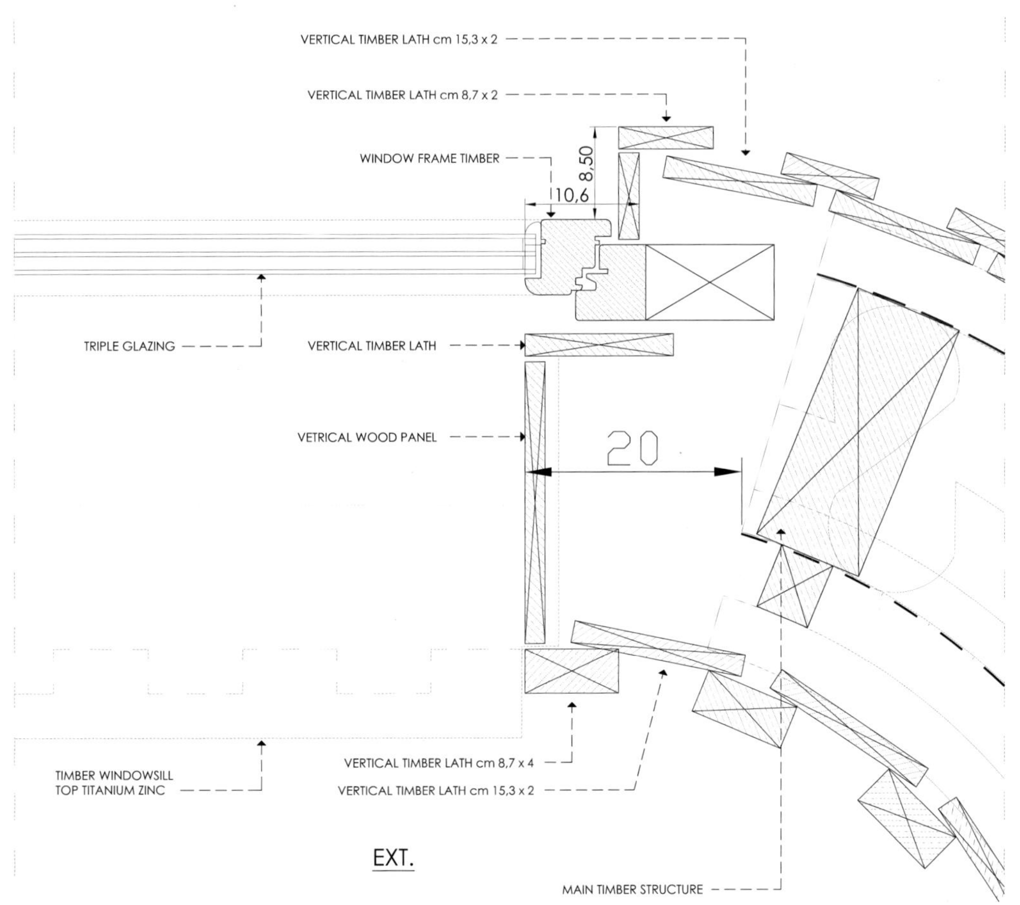

Detail - Horizontal window node

1.

2.

3.

4.

5.

6.

7.

Construction Process

© Alexandre Zweiger

© Alexandre Zweiger

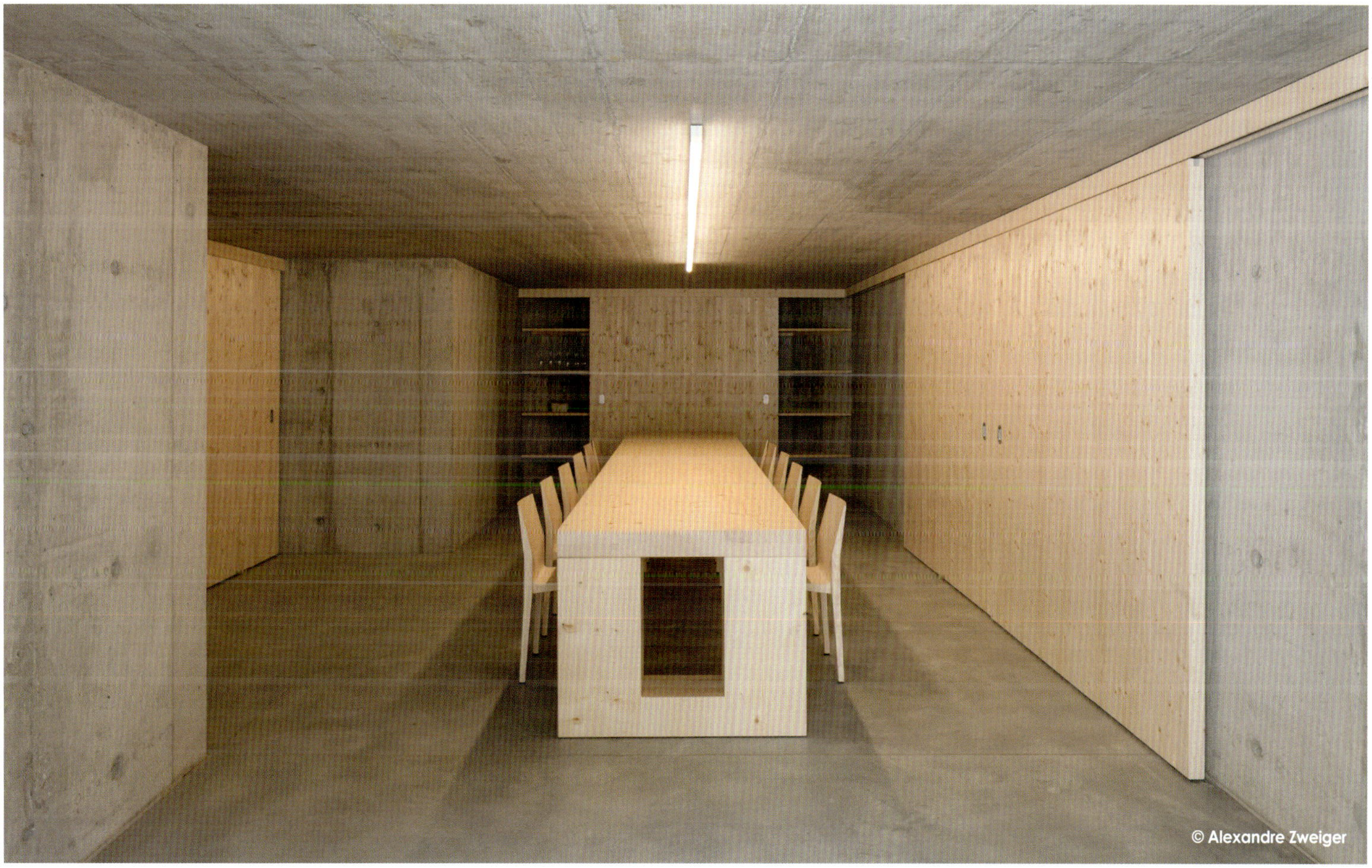

© Alexandre Zweiger

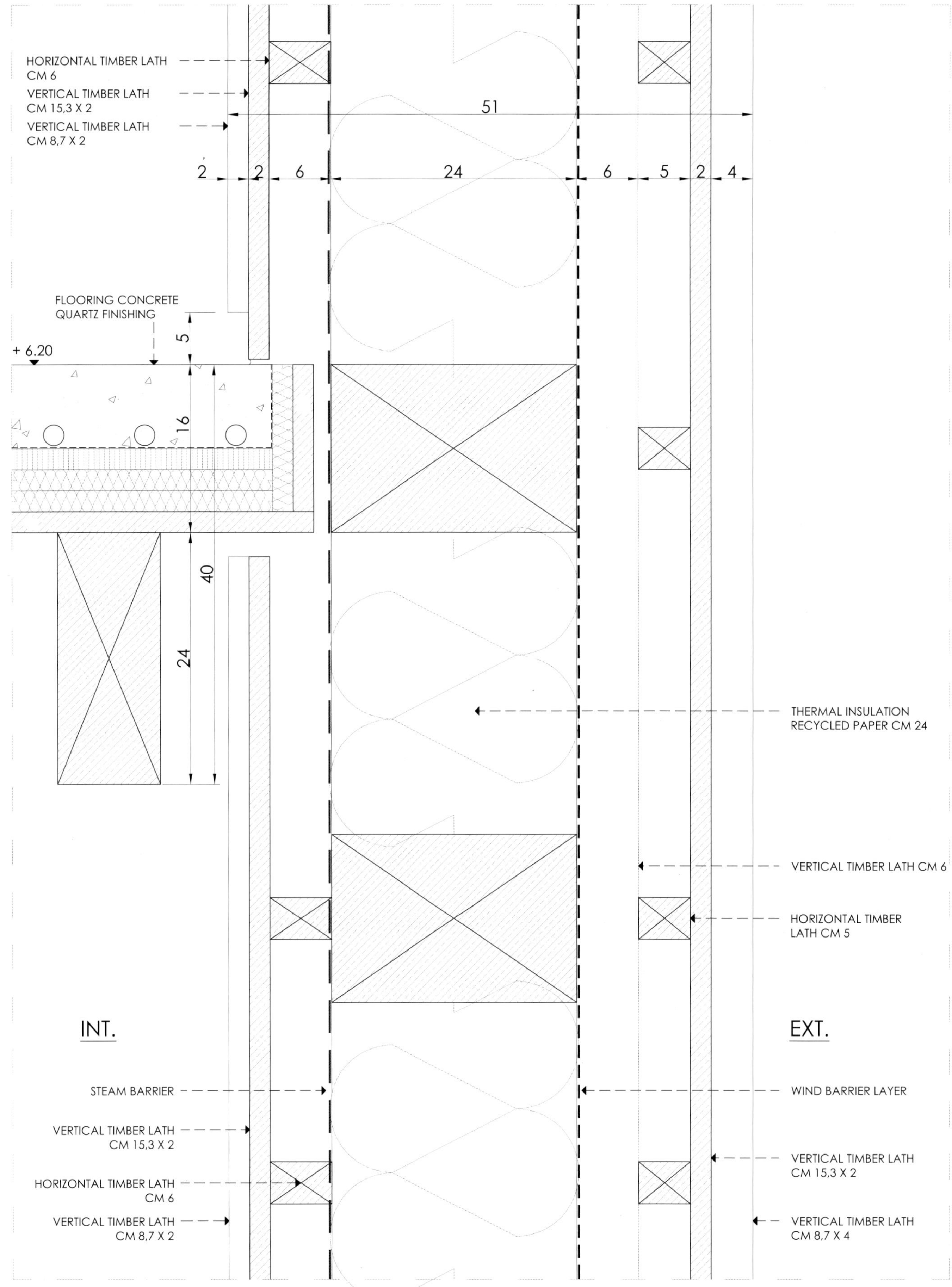

Detail - Floor to wall

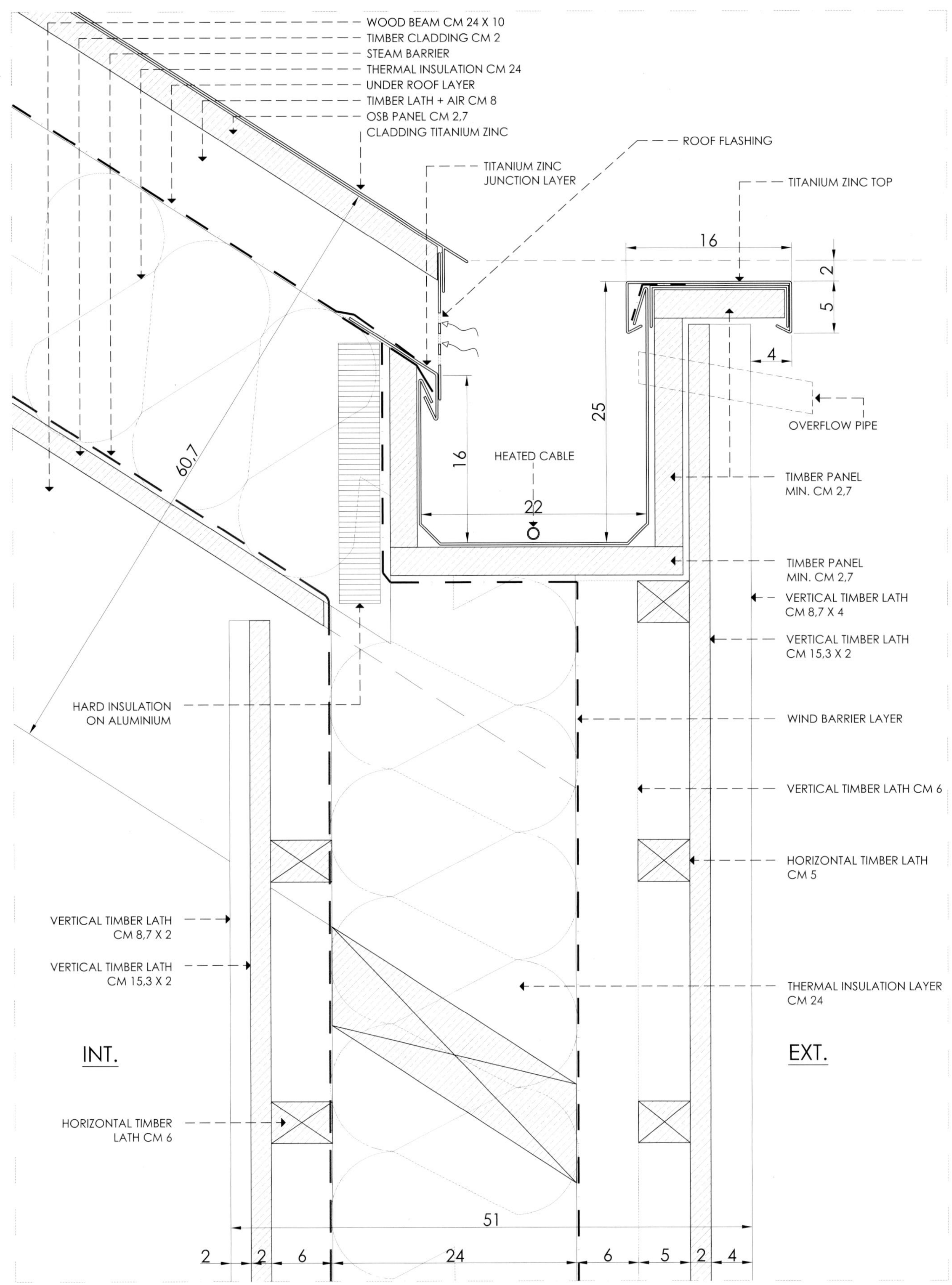
WOOD BEAM CM 24 X 10
TIMBER CLADDING CM 2
STEAM BARRIER
THERMAL INSULATION CM 24
UNDER ROOF LAYER
TIMBER LATH + AIR CM 8
OSB PANEL CM 2,7
CLADDING TITANIUM ZINC
ROOF FLASHING
TITANIUM ZINC
JUNCTION LAYER
TITANIUM ZINC TOP
16
2
5
4
OVERFLOW PIPE
25
16
HEATED CABLE
22
60,7
TIMBER PANEL
MIN. CM 2,7
TIMBER PANEL
MIN. CM 2,7
VERTICAL TIMBER LATH
CM 8,7 X 4
VERTICAL TIMBER LATH
CM 15,3 X 2
HARD INSULATION
ON ALUMINIUM
WIND BARRIER LAYER
VERTICAL TIMBER LATH CM 6
HORIZONTAL TIMBER LATH
CM 5
VERTICAL TIMBER LATH
CM 8,7 X 2
VERTICAL TIMBER LATH
CM 15,3 X 2
THERMAL INSULATION LAYER
CM 24
INT.
EXT.
HORIZONTAL TIMBER
LATH CM 6
51
2
2
6
24
6
5
2
4

Roof Detail

© Alexandre Zweiger

T noie, Toyota Aichi, Japan

Katsutoshi Sasaki + Associates

© Katsutoshi Sasaki + Associates

© Katsutoshi Sasaki + Associates

© Katsutoshi Sasaki + Associates

© Katsutoshi Sasaki + Associates

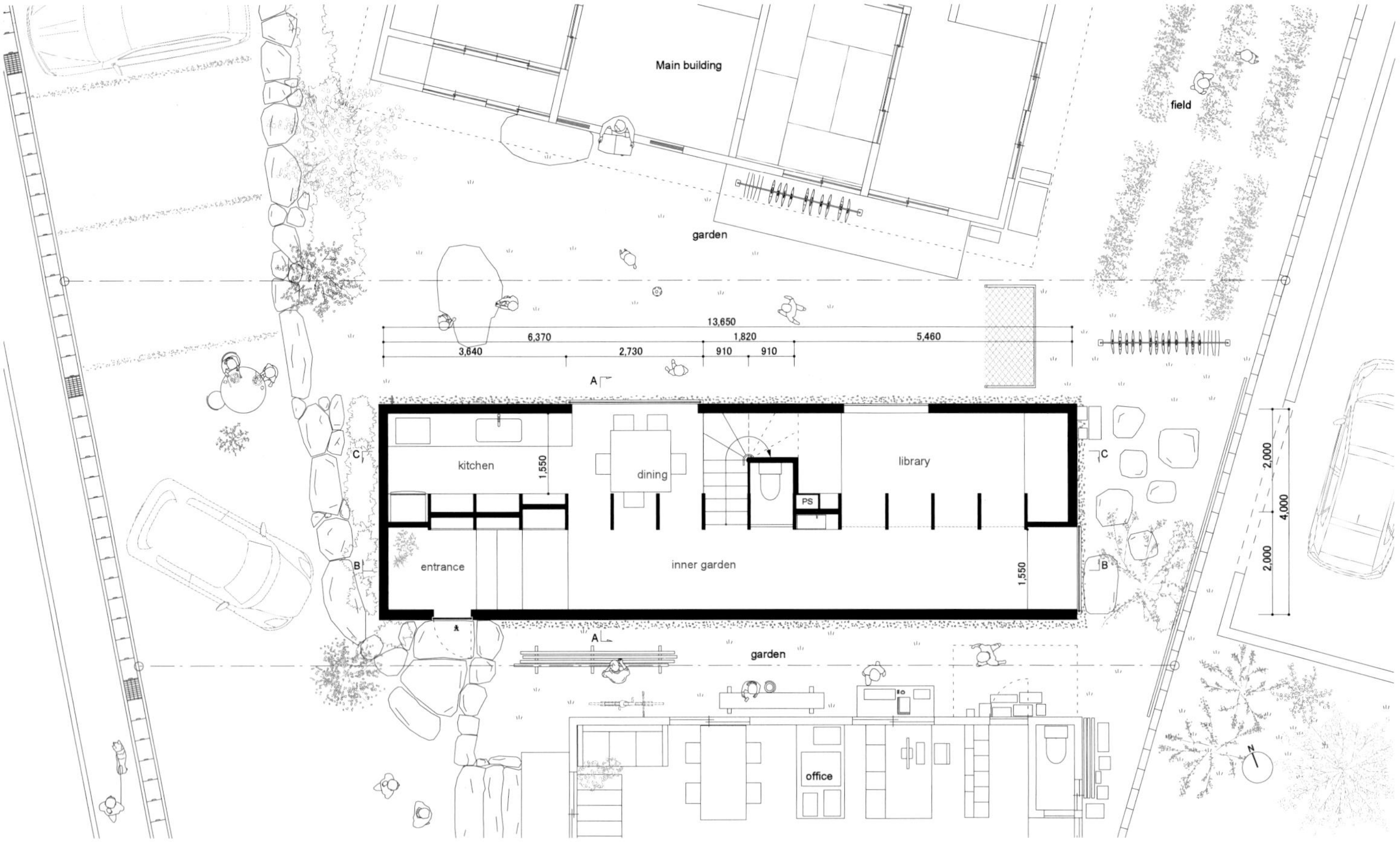

Main building
field
garden
13,650
6,370
1,820
5,460
3,640
2,730
910
910
kitchen
dining
library
PS
entrance
inner garden
1,550
1,550
2,000
4,000
2,000
garden
office

Ground Floor

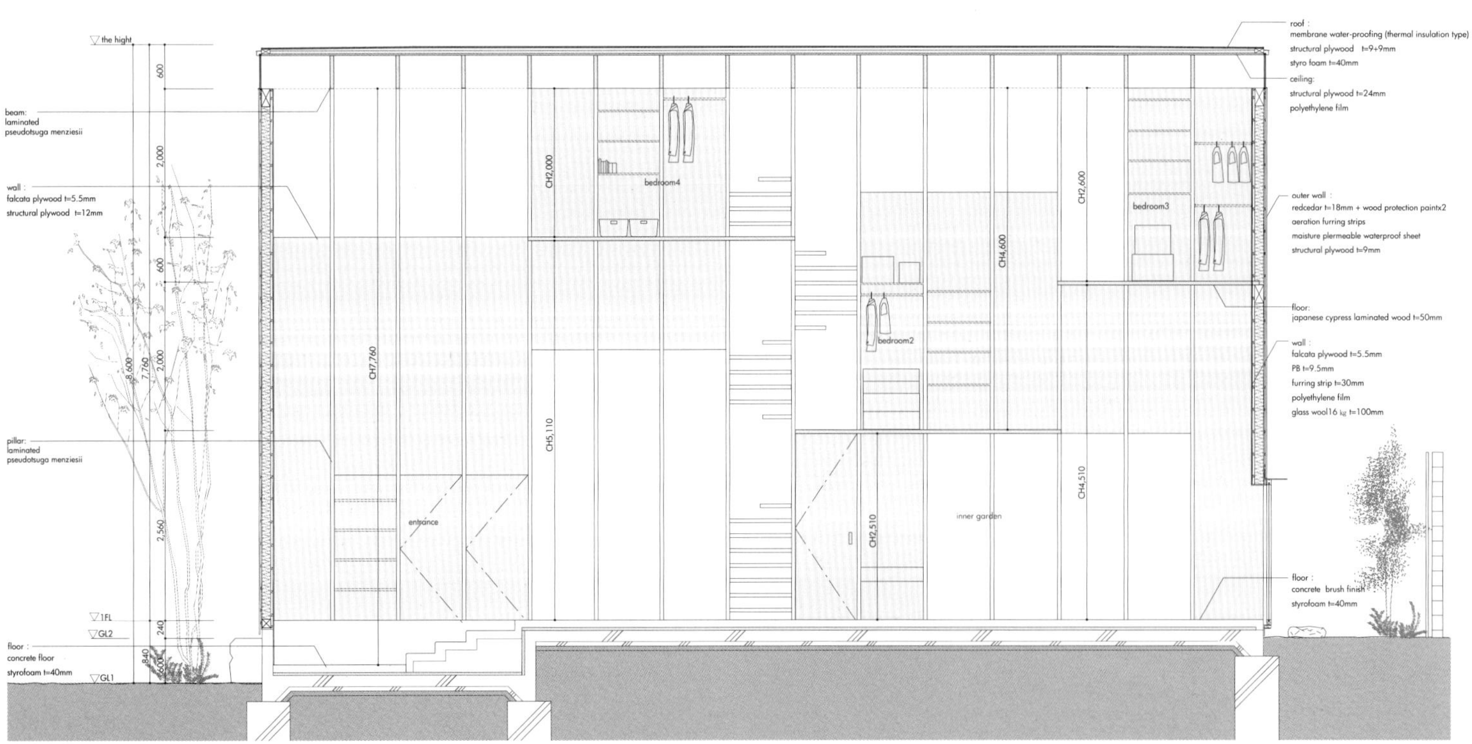

Section B

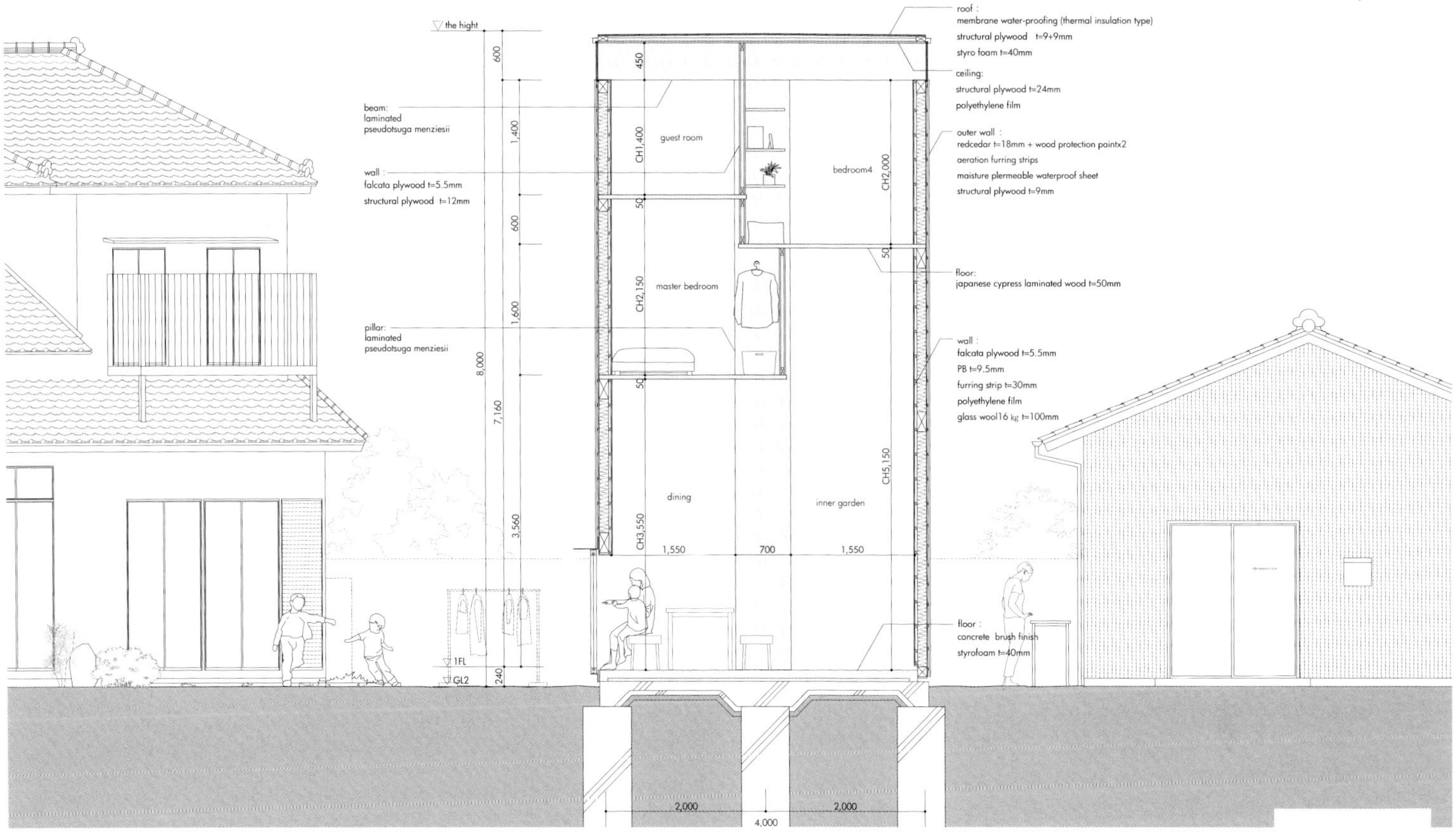

the hight

roof :
membrane water-proofing (thermal insulation type)
structural plywood t=9+9mm
styro foam t=40mm

ceiling:
structural plywood t=24mm
polyethylene film

beam:
laminated
pseudotsuga menziesii

wall :
falcata plywood t=5.5mm
structural plywood t=12mm

outer wall :
redcedar t=18mm + wood protection paintx2
aeration furring strips
moisture plermeable waterproof sheet
structural plywood t=9mm

guest room

bedroom4

master bedroom

floor:
japanese cypress laminated wood t=50mm

pillar:
laminated
pseudotsuga menziesii

wall :
falcata plywood t=5.5mm
PB t=9.5mm
furring strip t=30mm
polyethylene film
glass wool16 kg t=100mm

dining

inner garden

floor :
concrete brush finish
styrofoam t=40mm

1FL
GL2

600 | 1,400 | 600 | 1,600 | 3,560 | 240 | 8,000 | 7,160

450 | CH1,400 | 50 | CH2,150 | 50 | CH3,550 | CH2,000 | 50 | CH5,150

1,550 | 700 | 1,550

2,000 | 2,000 | 4,000

Sectional Detail

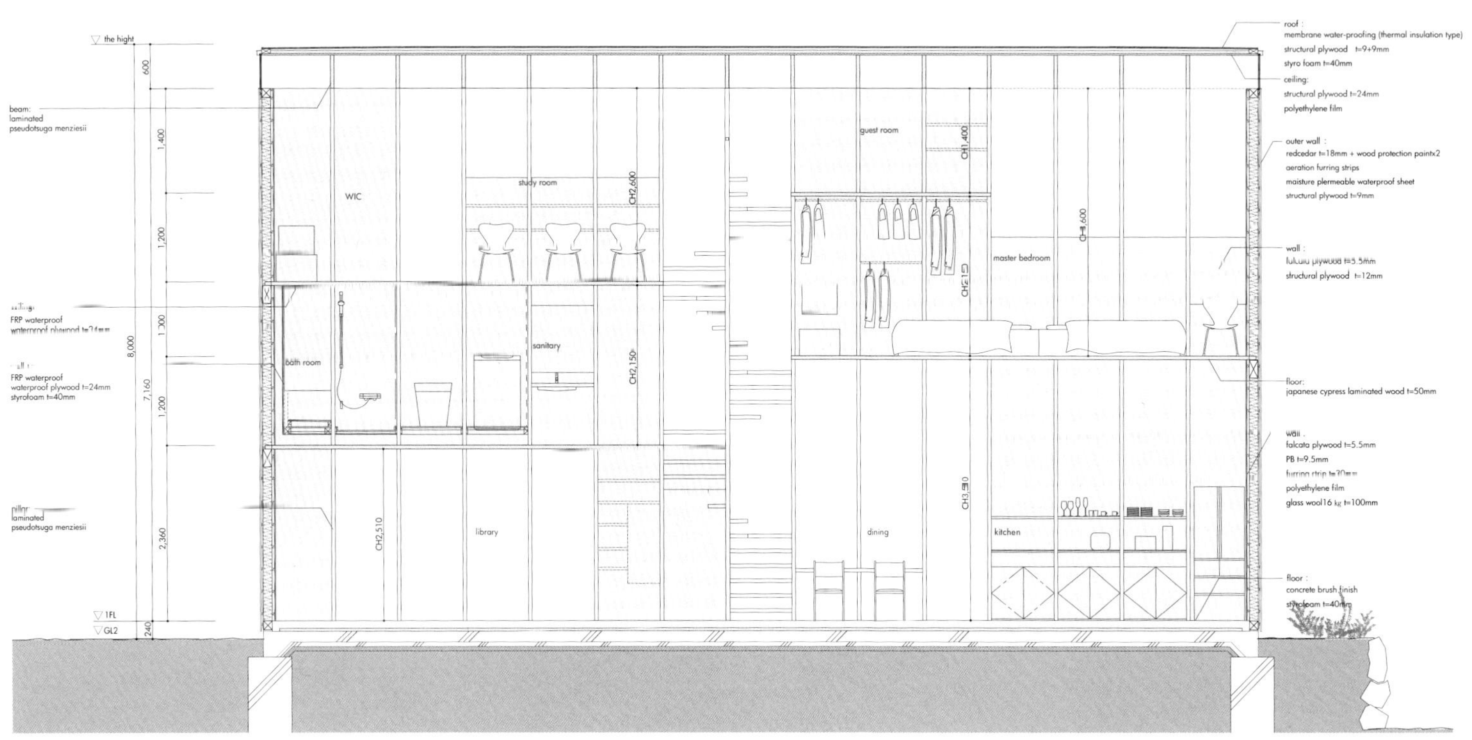

Section C

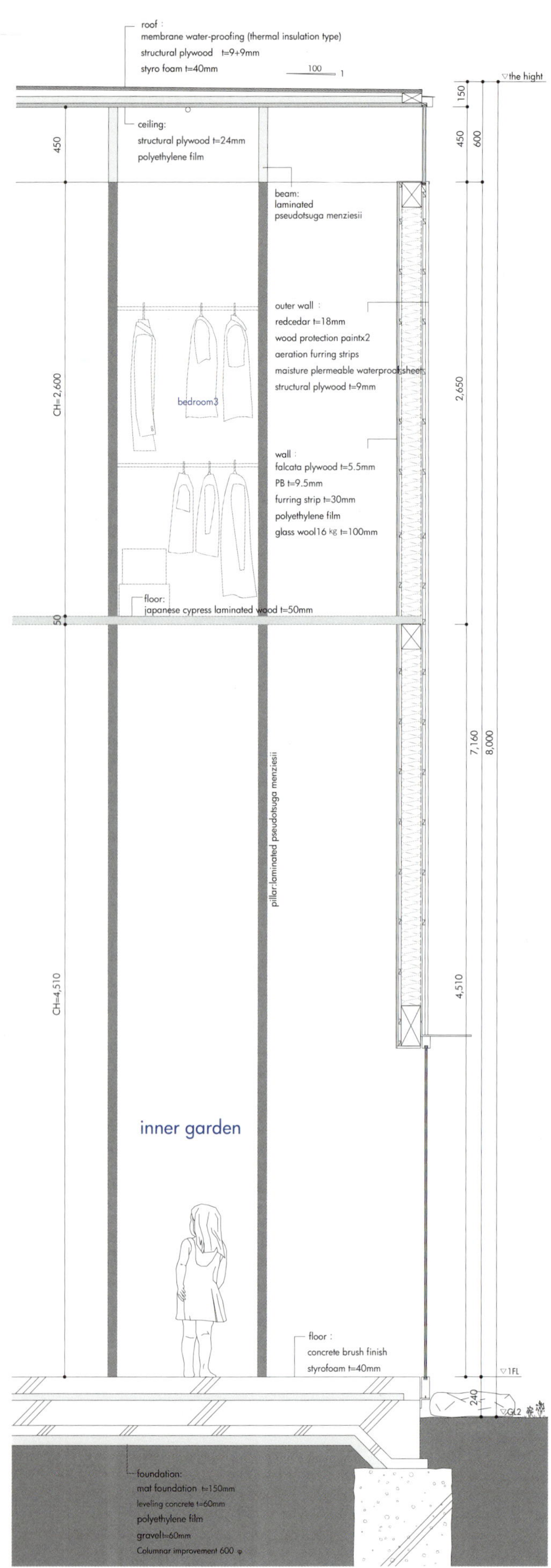

Section Detail

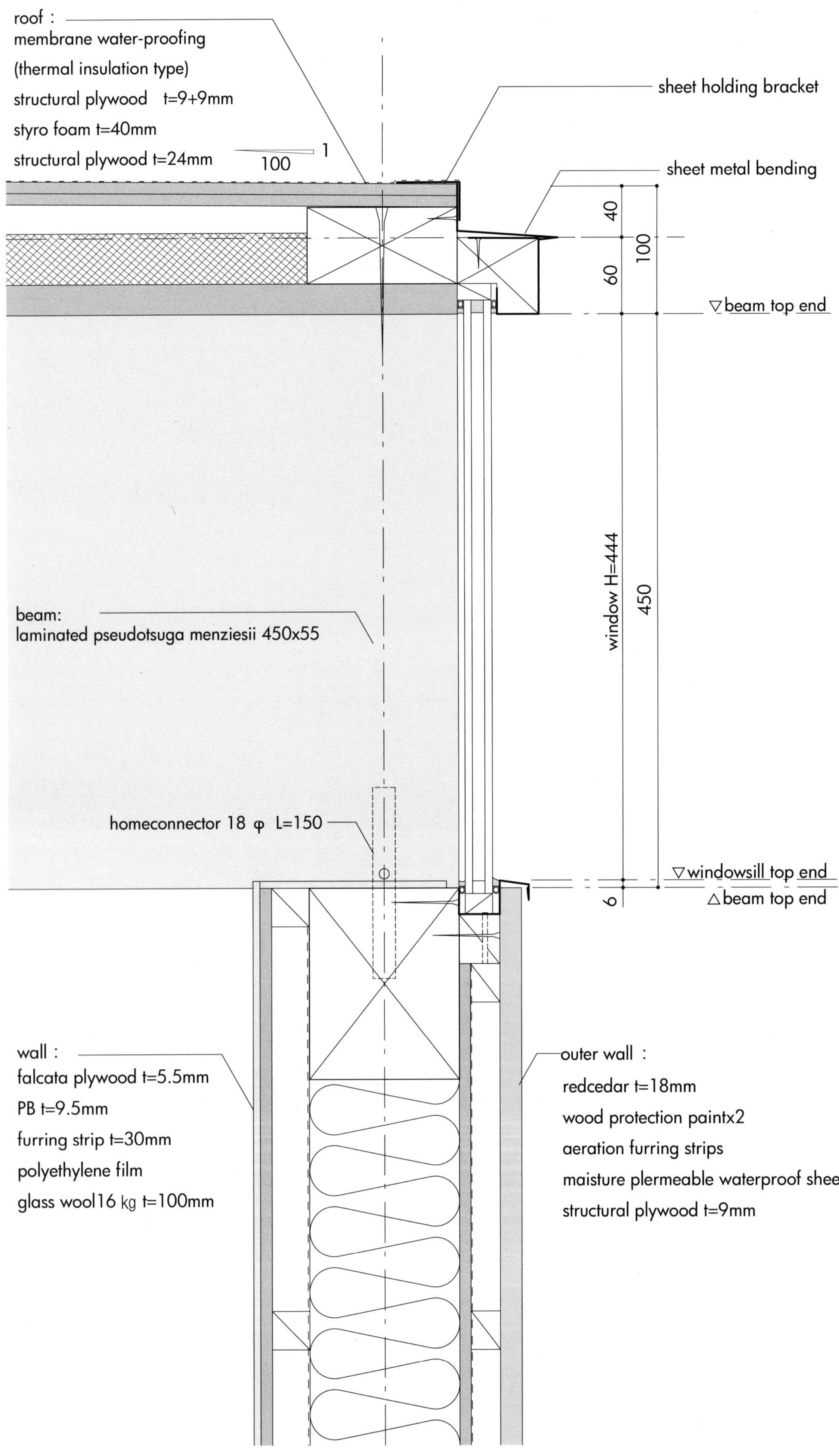

Window Detail

佐々木勝敏建築設計事務所

unou, Toyota Aichi, Japan
Katsutoshi Sasaki + Associates

© Alexandre Zweiger

garage

Site Plan

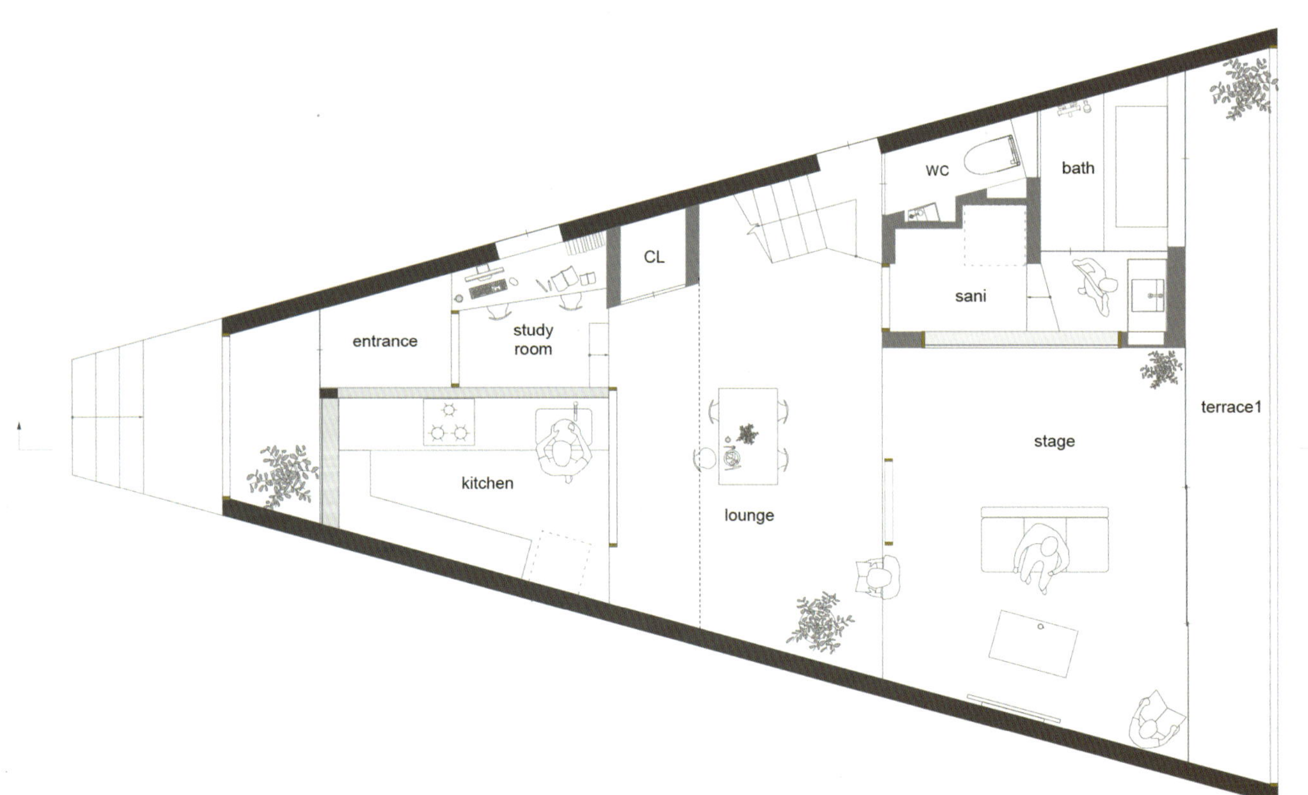

First Floor

© Katsutoshi Sasaki + Associates

© Katsutoshi Sasaki + Associates

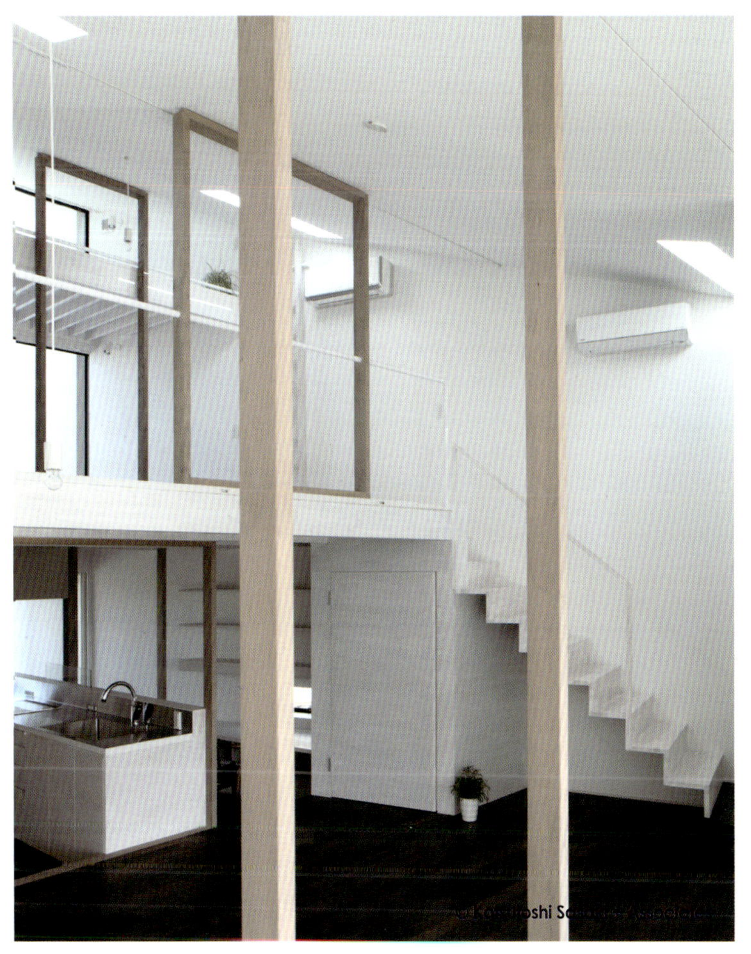

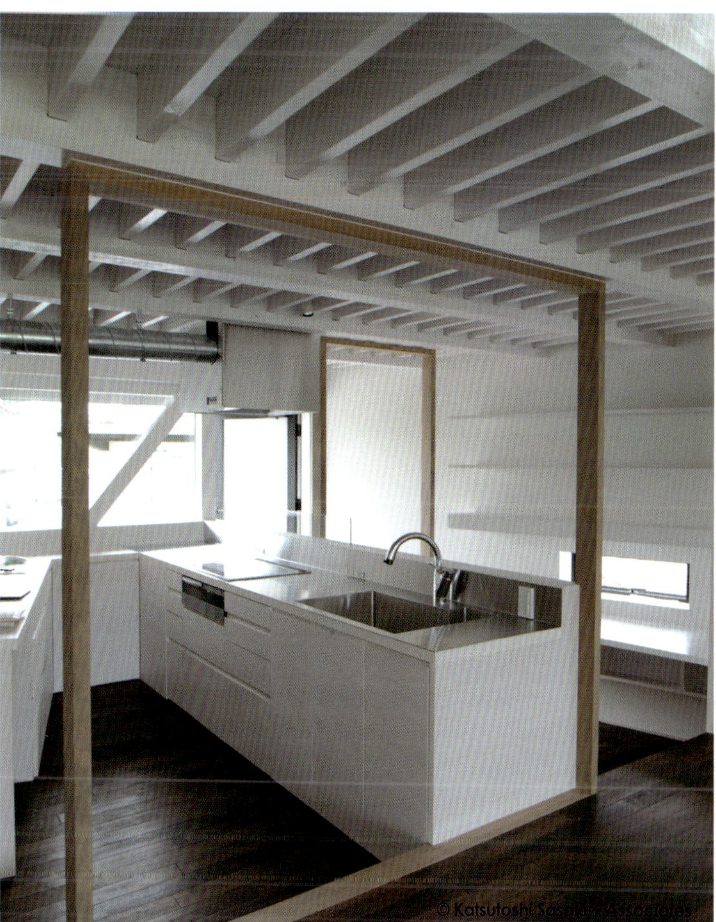

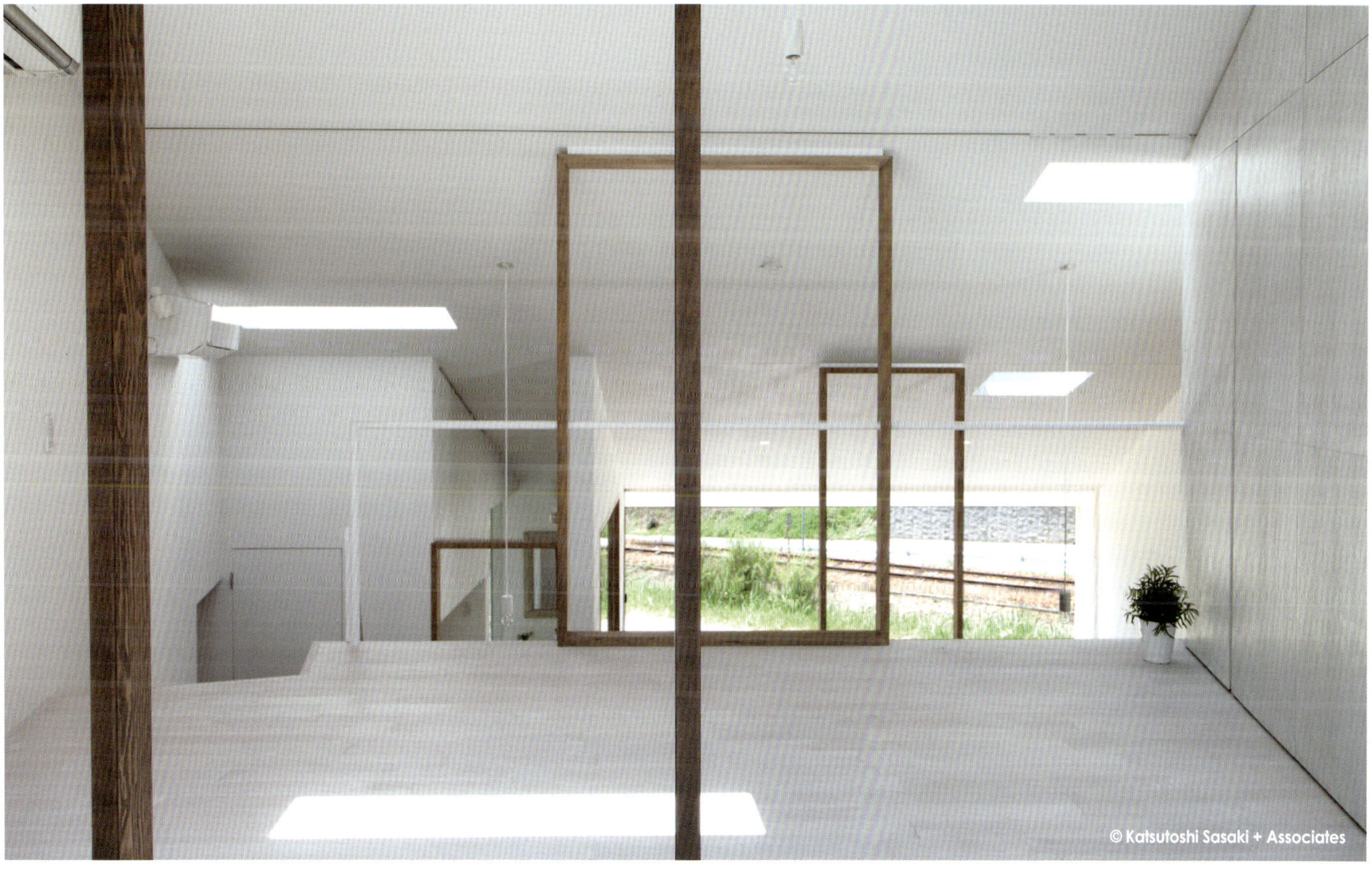
© Katsutoshi Sasaki + Associates

© Katsutoshi Sasaki + Associates

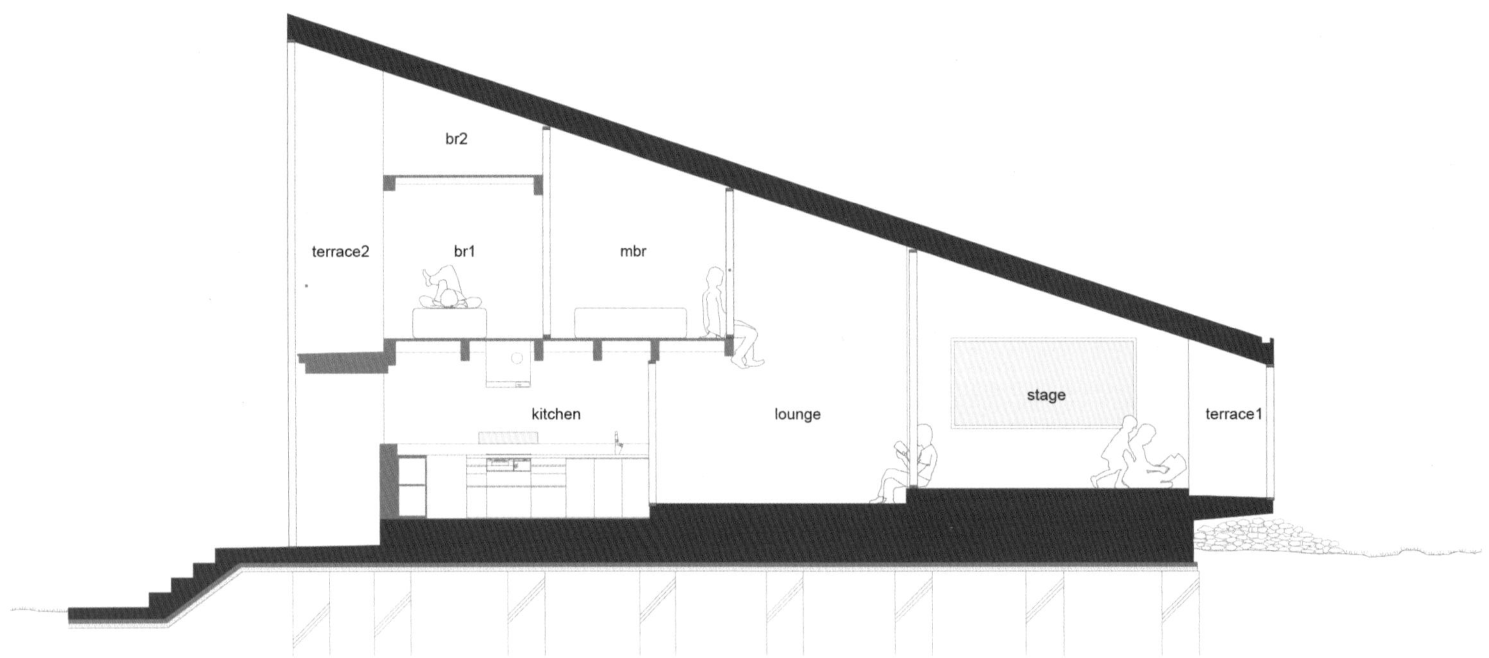

Section

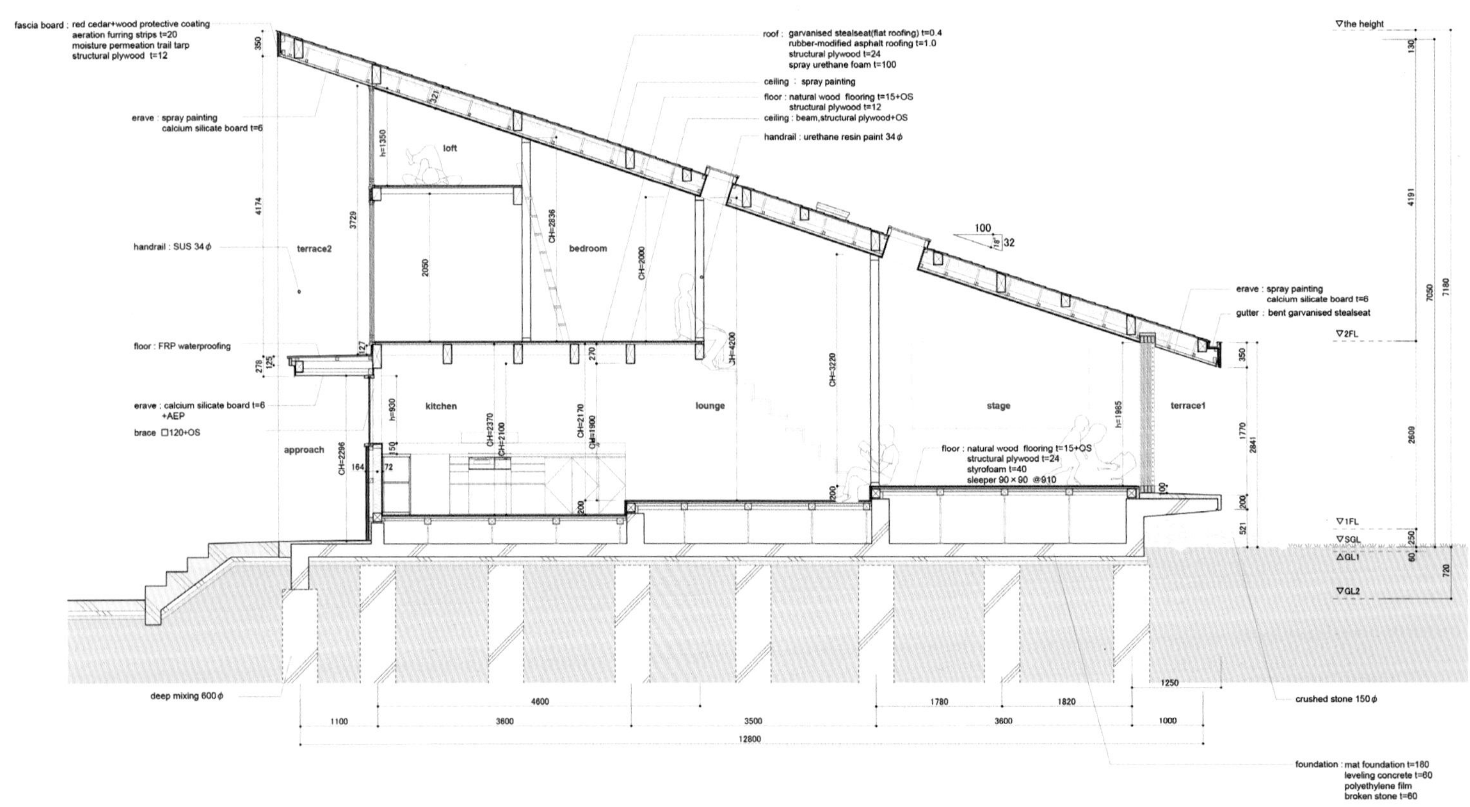

First Floor

© Katsutoshi Sasaki + Associates

V-House, Oslo, Norway

SPACEGROUP

WOOD+STEEL HOUSE, Yeoju, South Korea

ecosistema urbano

© EMILIO P. DOIZTUA

Process Diagram

© EMILIO P. DOIZTUA

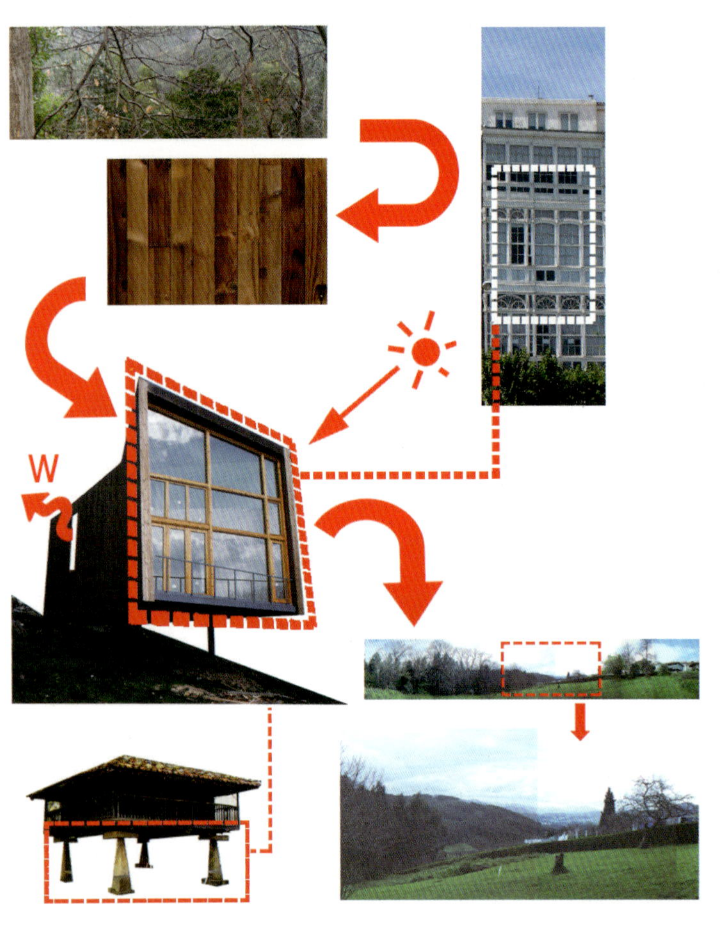
W

© EMILIO P. DOIZTUA

© EMILIO P. DOIZTUA

© EMILIO P. DOIZTUA

House in Chau Doc, An Giang, Viet Nam

NISHIZAWA ARCHITECTS

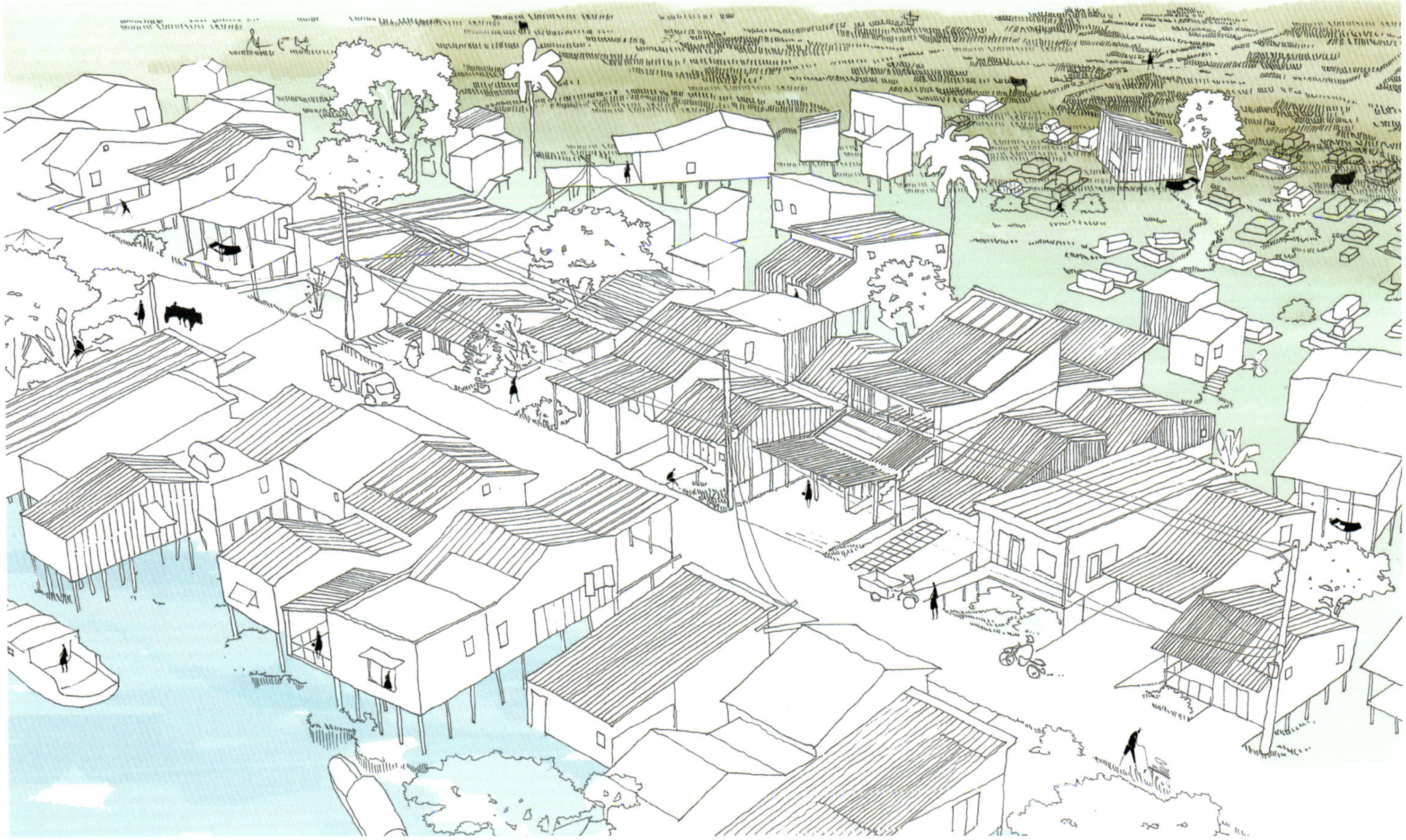

Sketch

Site Plan

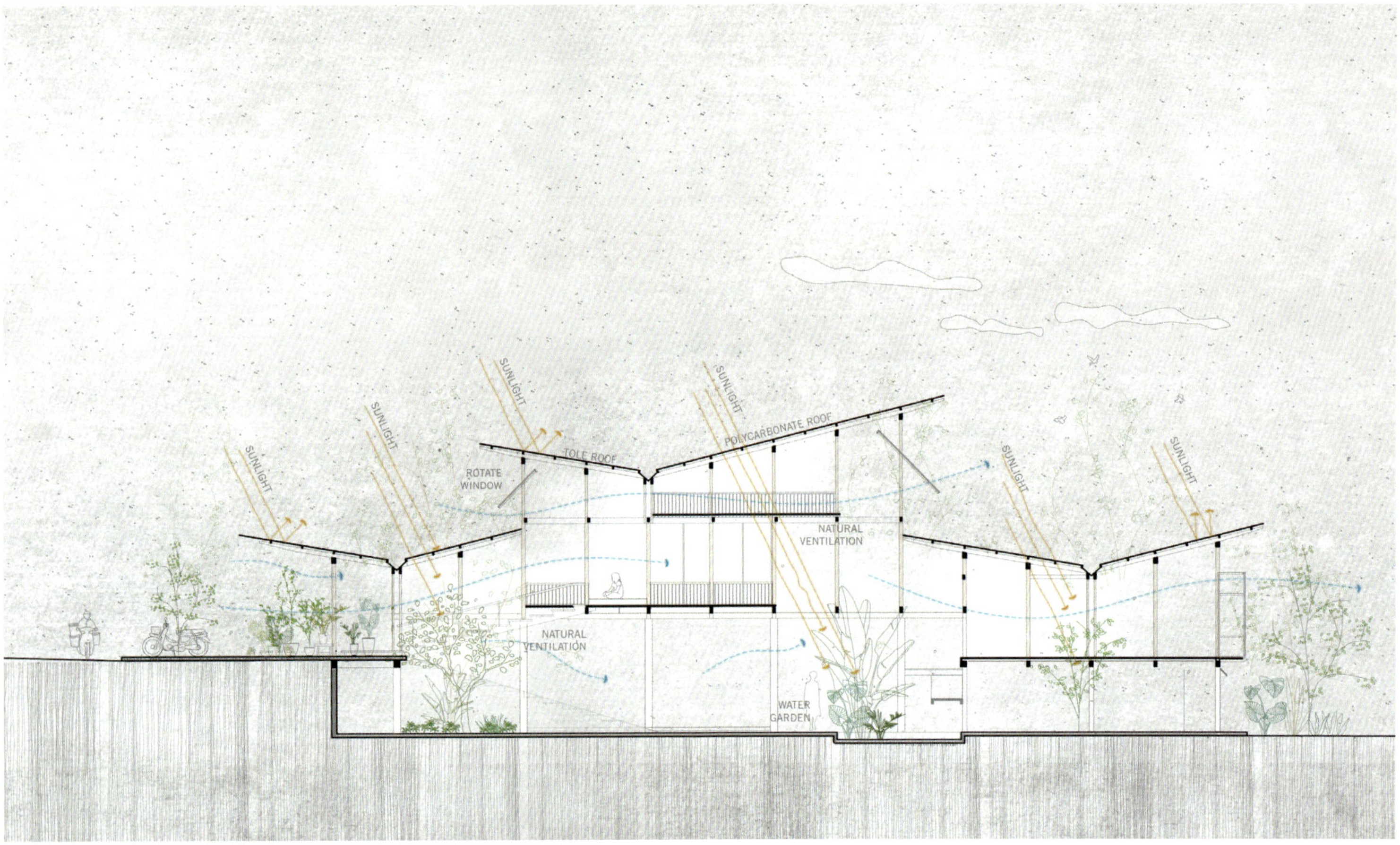
SUNLIGHT
SUNLIGHT
SUNLIGHT
SUNLIGHT
SUNLIGHT
SUNLIGHT
TOLE ROOF
POLYCARBONATE ROOF
ROTATE
WINDOW
NATURAL
VENTILATION
NATURAL
VENTILATION
WATER
GARDEN

Section A

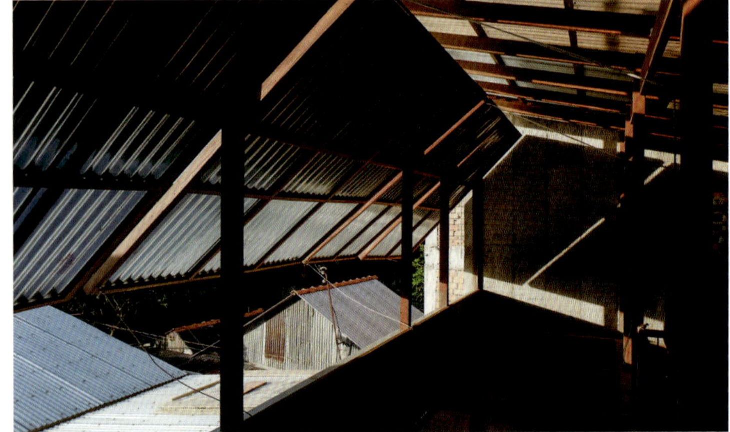

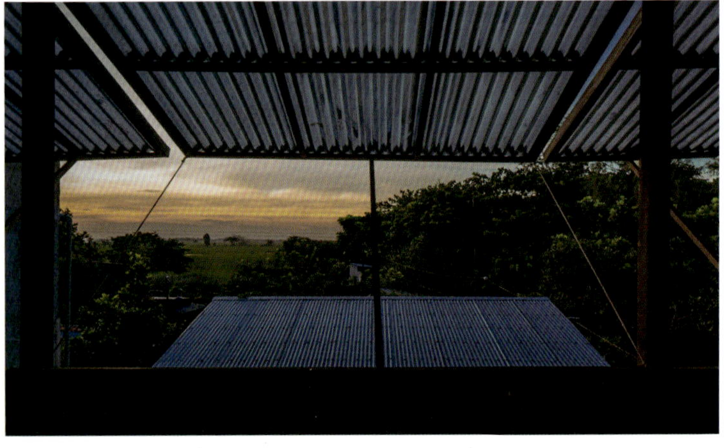

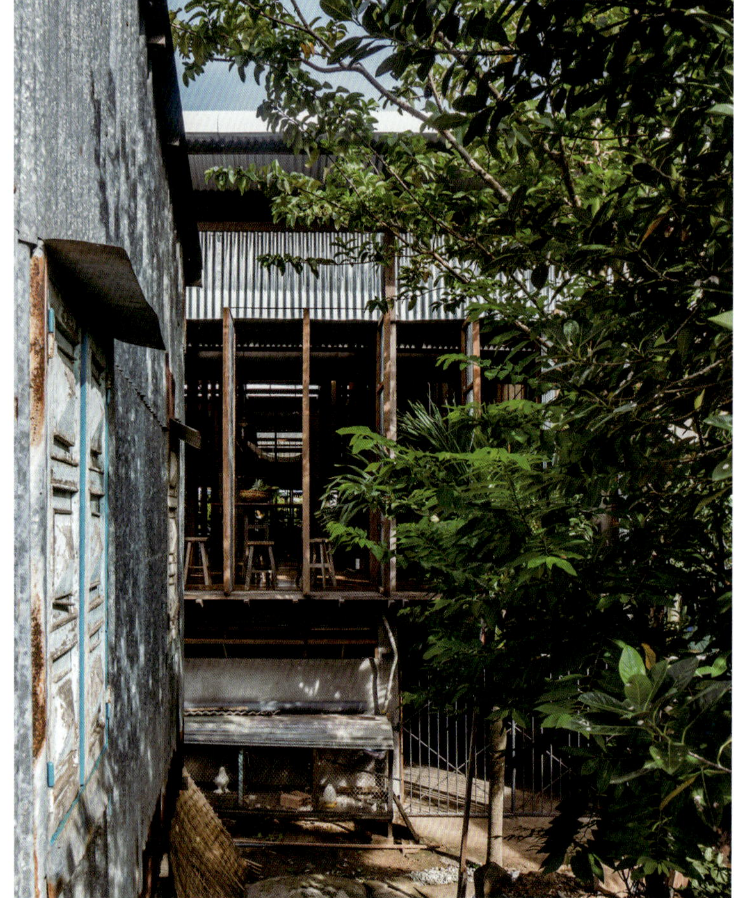

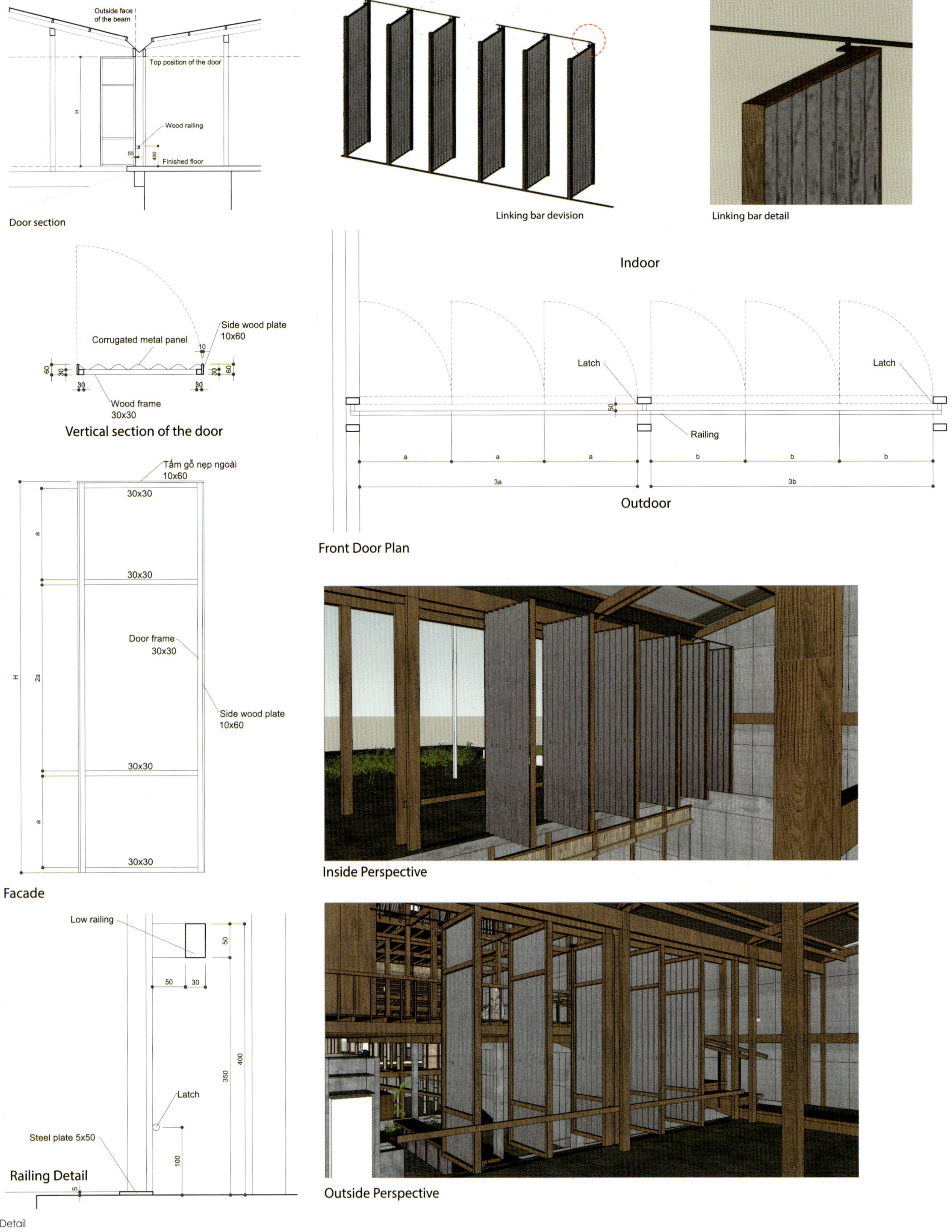

Door section

Linking bar devision

Linking bar detail

Front Door Plan

Vertical section of the door

Facade

Inside Perspective

Railing Detail

Outside Perspective

Detail

Patio Villas, Groningen, the Netherlands

Casanova+Hernandez Architects

© Van Der Vlugt

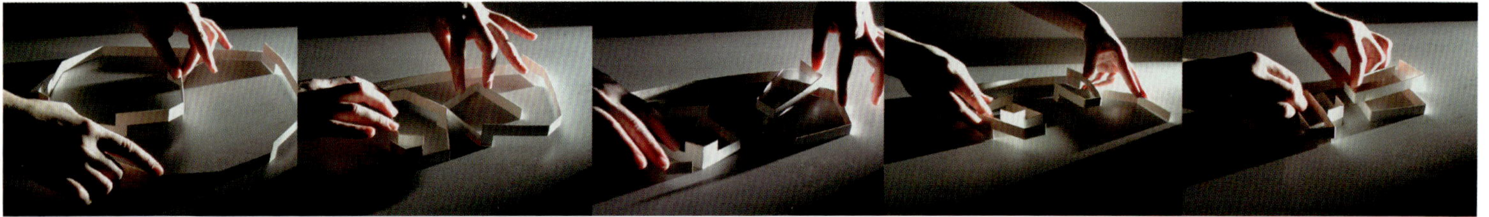

Conceptual Study

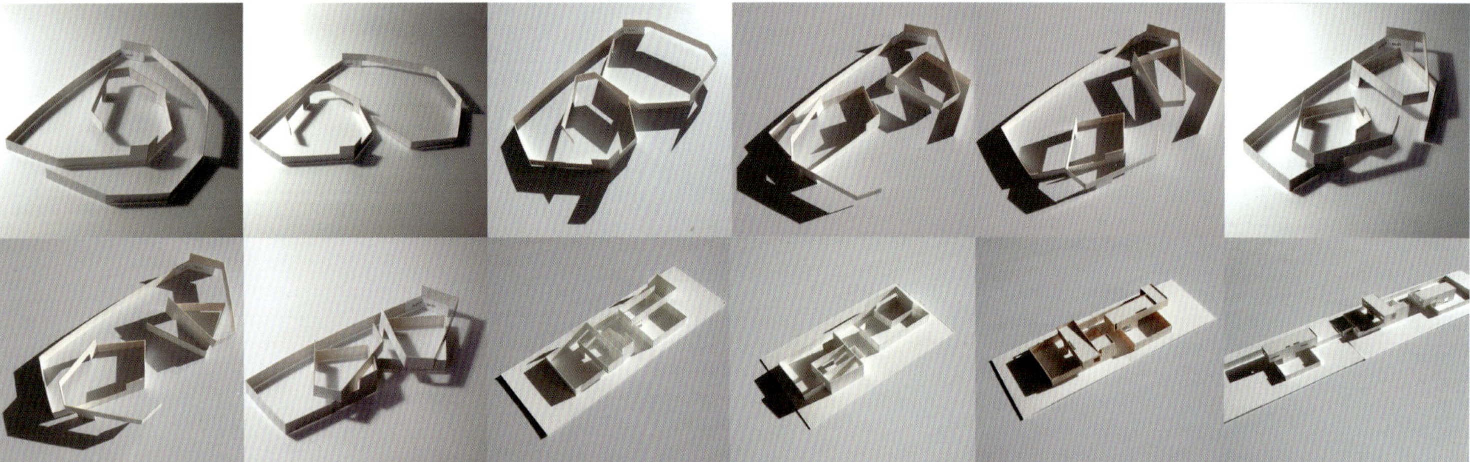

Morphological Study

Design Strategy

Parameters
: Design flexibility_Unique identity_Spatial diversity

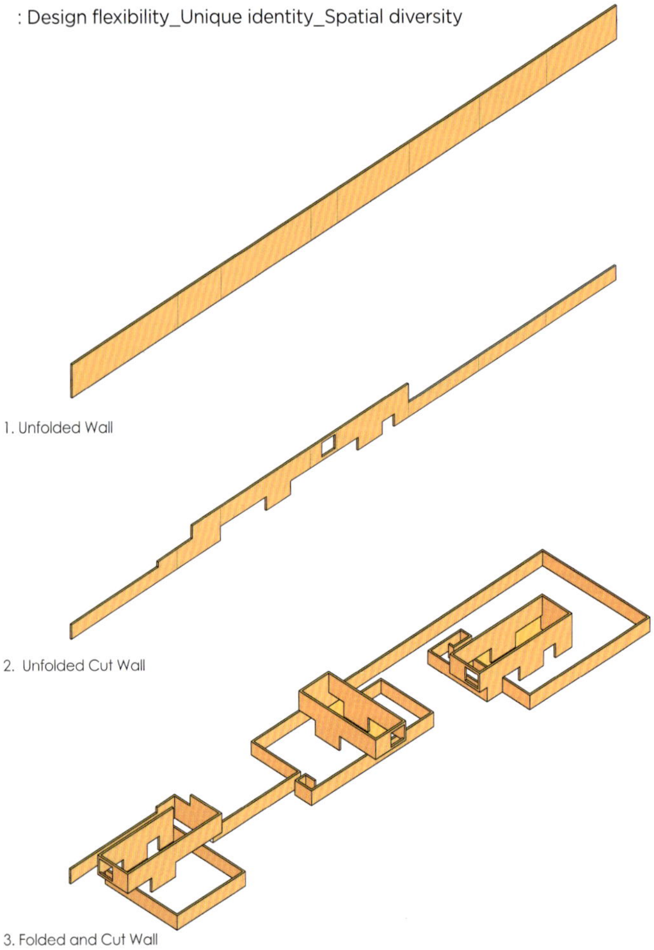

1. Unfolded Wall

2. Unfolded Cut Wall

3. Folded and Cut Wall

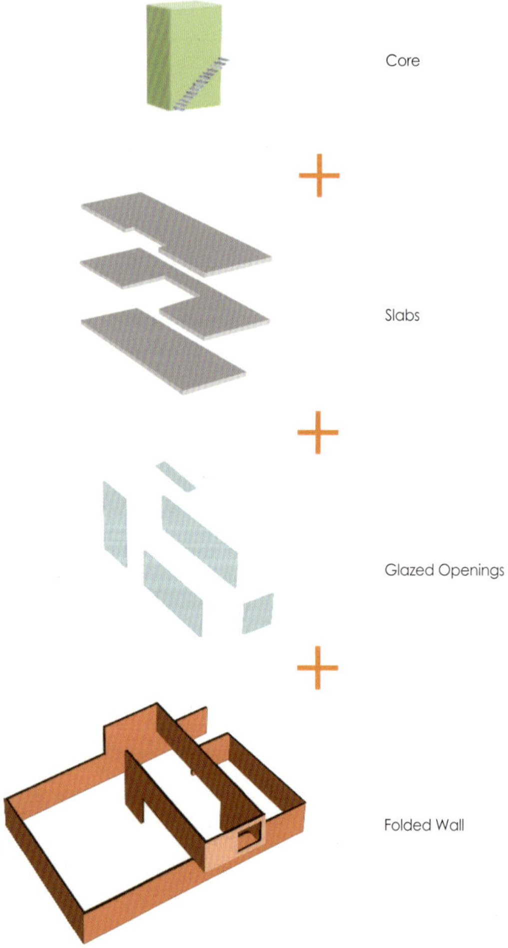

Core

+

Slabs

+

Glazed Openings

+

Folded Wall

© John Lewis Marshall

© John Lewis Marshall

01
02
03
04
05
06
07
08
09
10
11

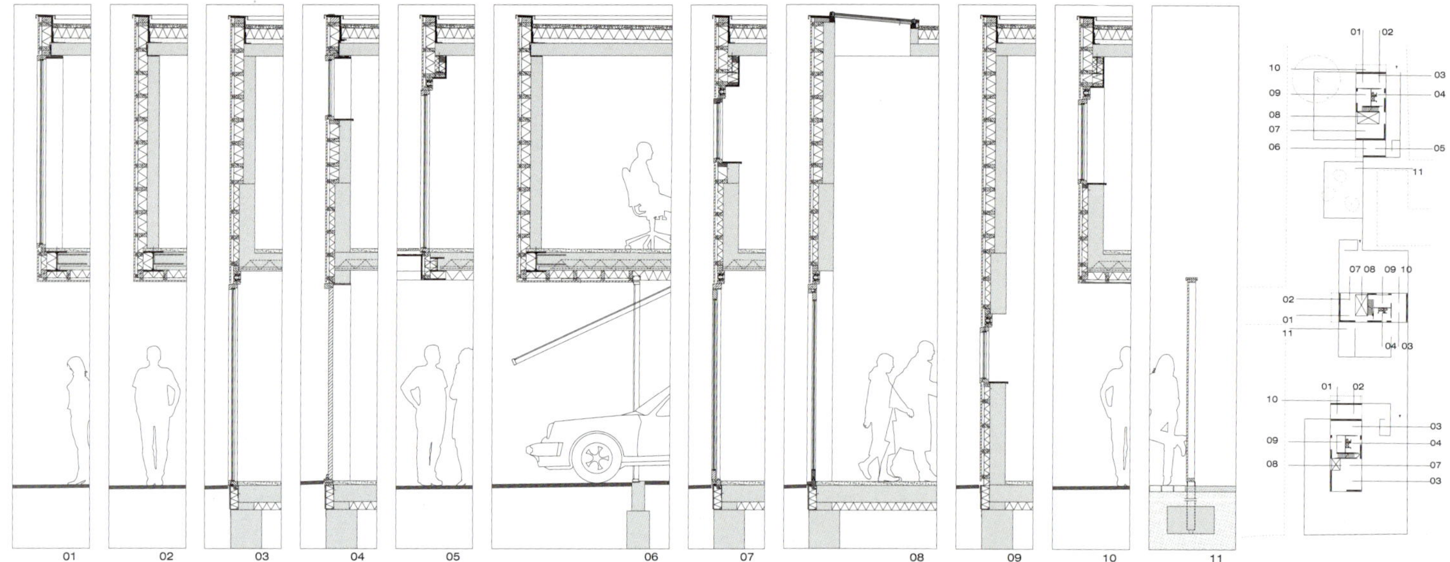

Catalogue of Constructive Solutions

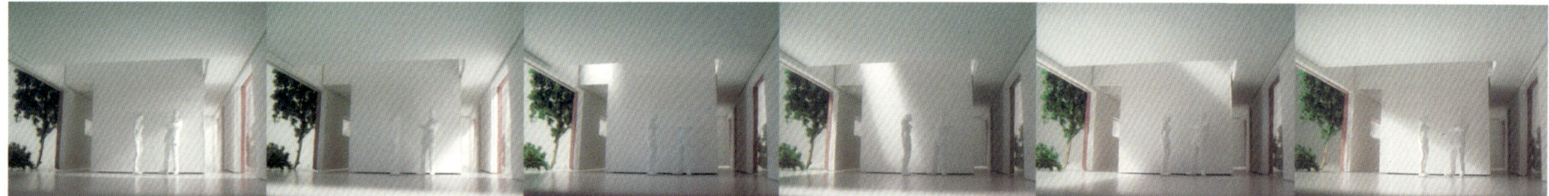

Interior Space Study

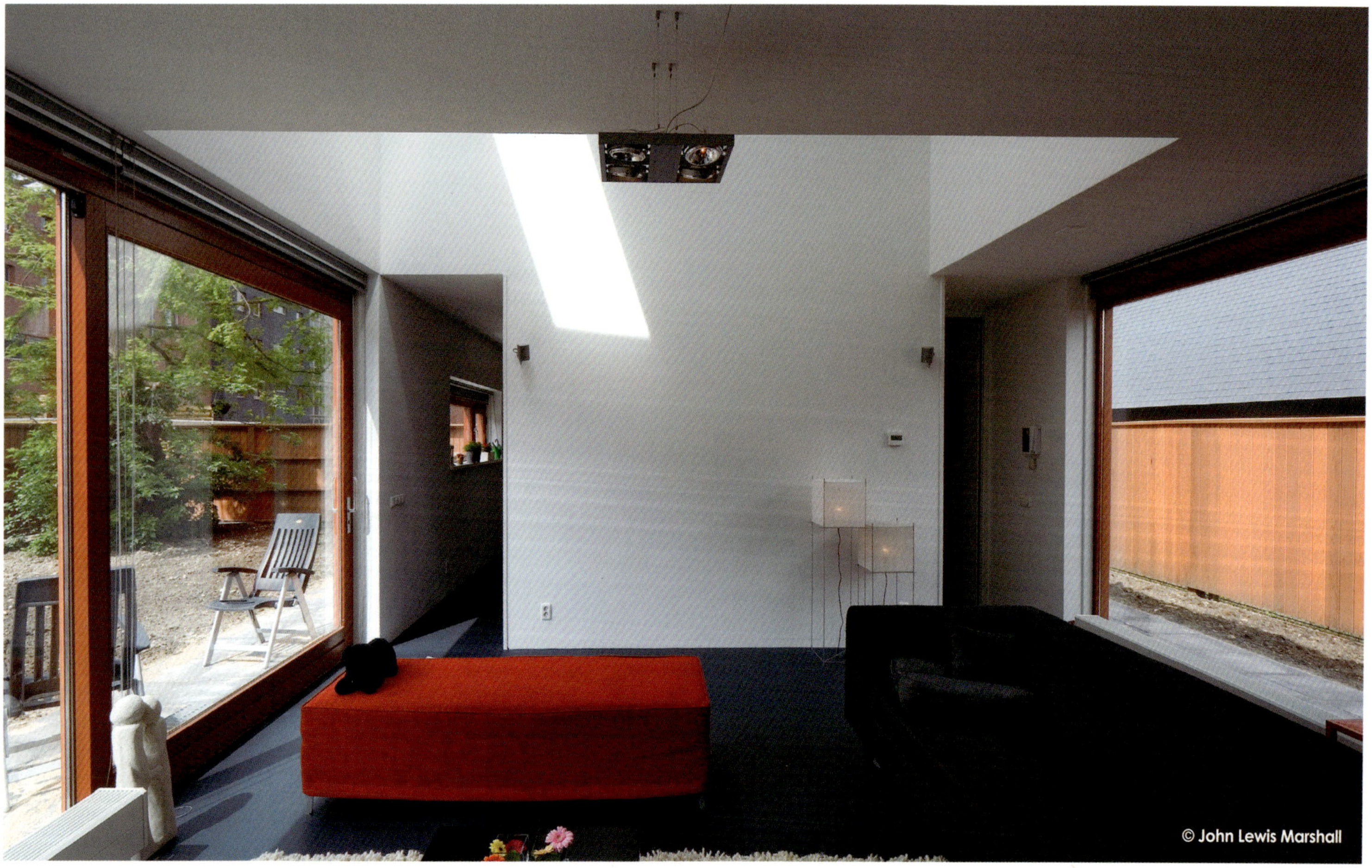

© John Lewis Marshall

© John Lewis Marshall

© Christian Richters

Shopping Roof Apartments, Bohinj, Slovenia

OFIS arhitekti

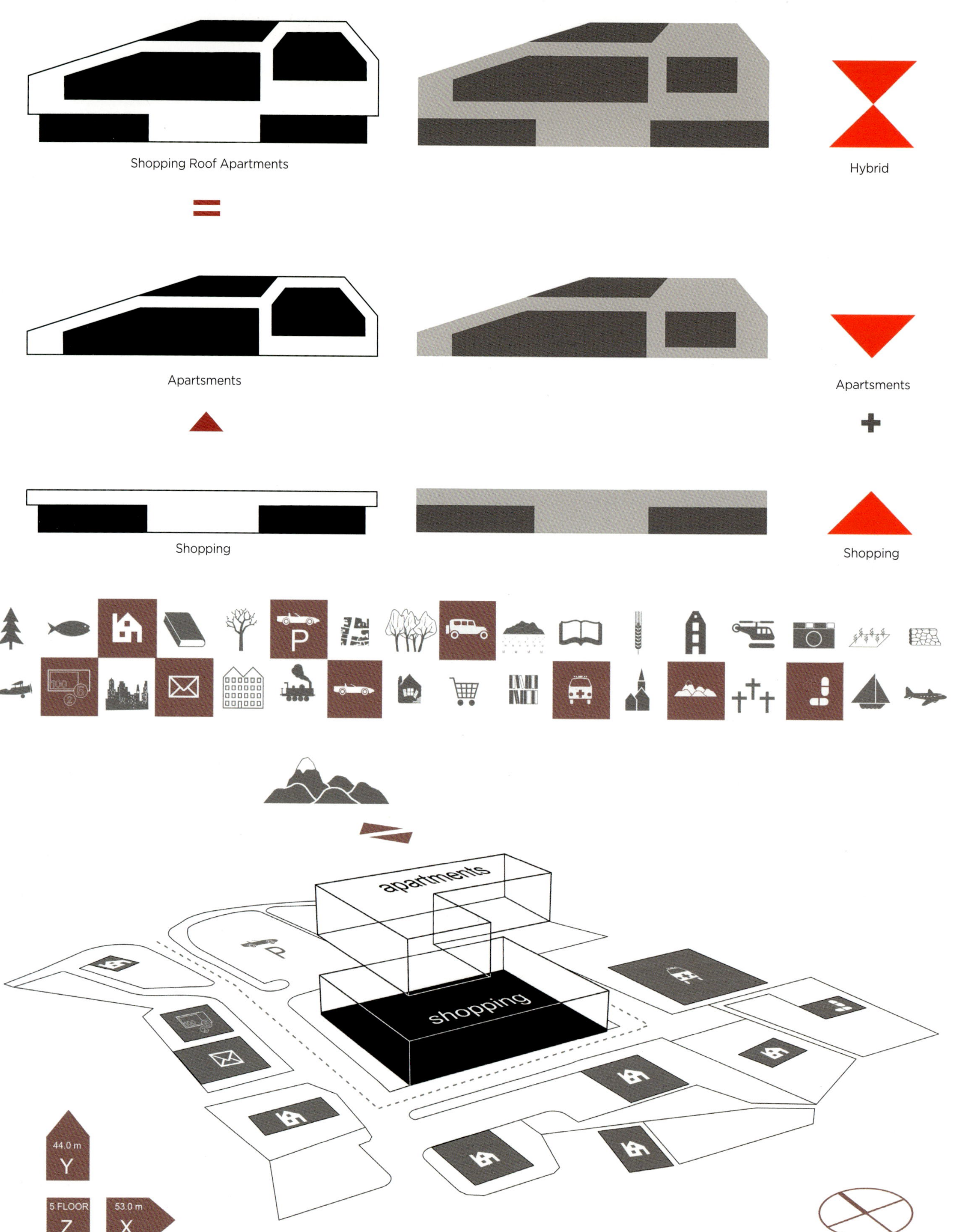
Shopping Roof Apartments
Hybrid
Apartments
Apartments
Shopping
Shopping
P
100
apartments
shopping
P
44.0 m
Y
5 FLOOR
Z
53.0 m
X

© Tomaz Gregoric

© Tomaz Gregoric

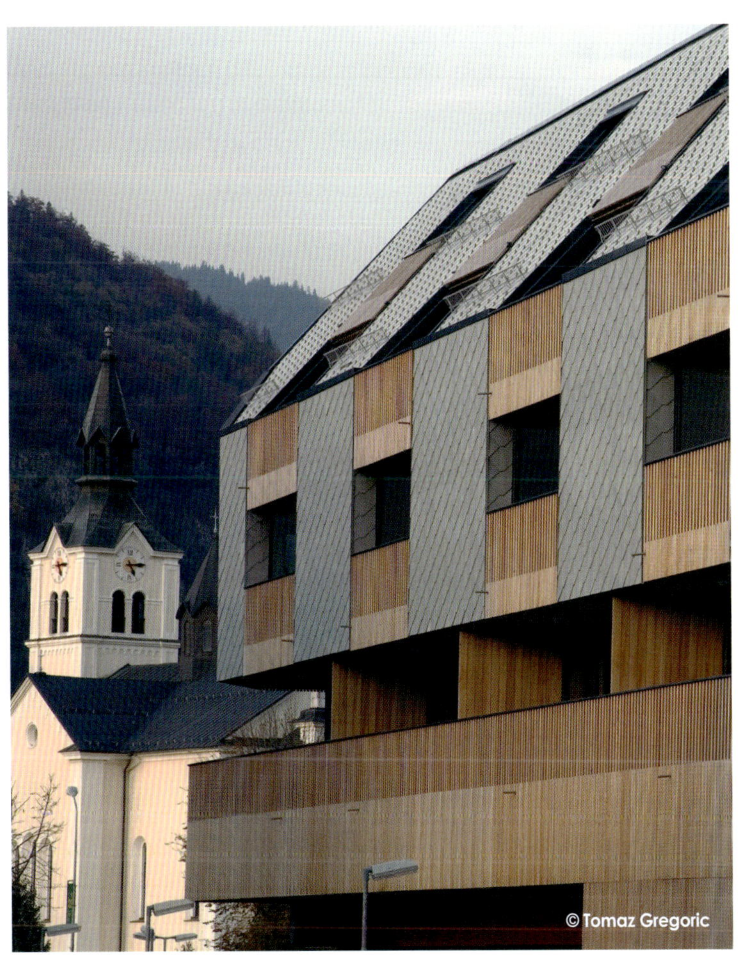

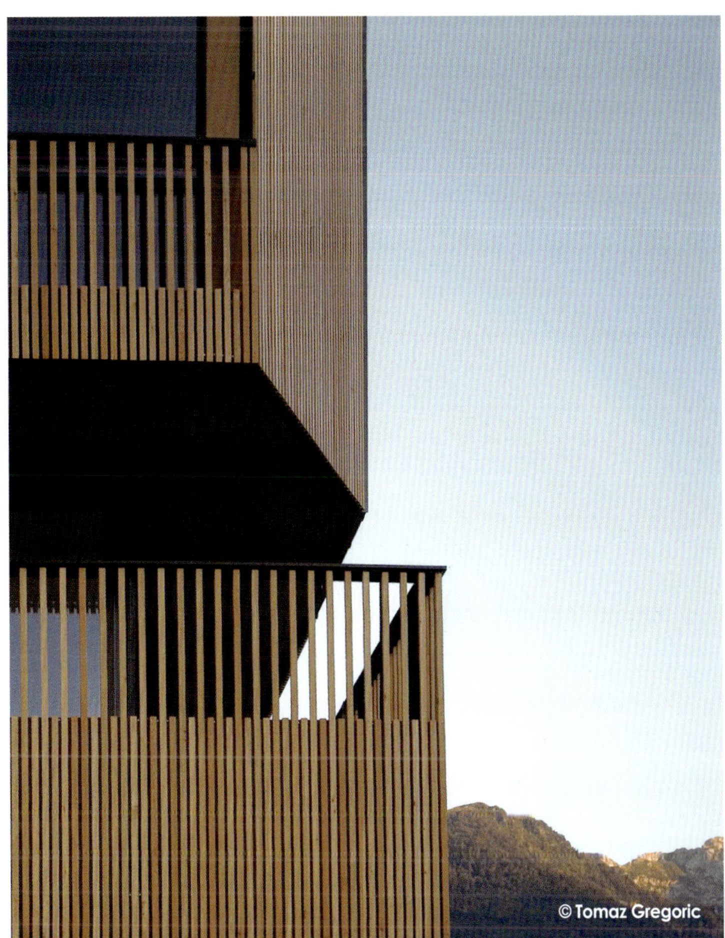

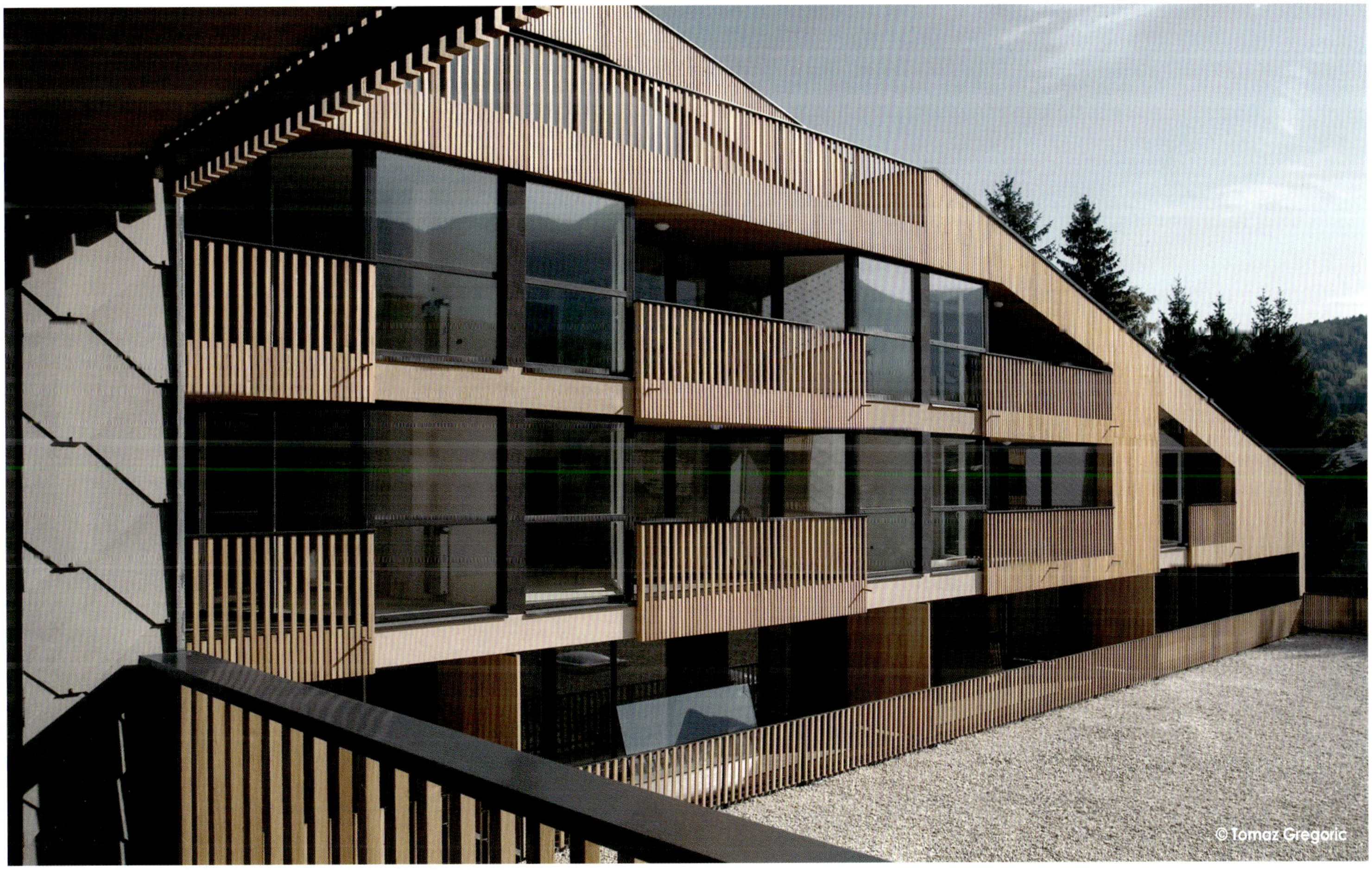
© Tomaz Gregoric

North Elevation

East Elevation

South Elevation

West Elevation

Fibre-Cement Plate | Wooden Laths | High Pressure Laminate Plates | Plaster (Granulate 1.5cm) | Glass

0 5 10m

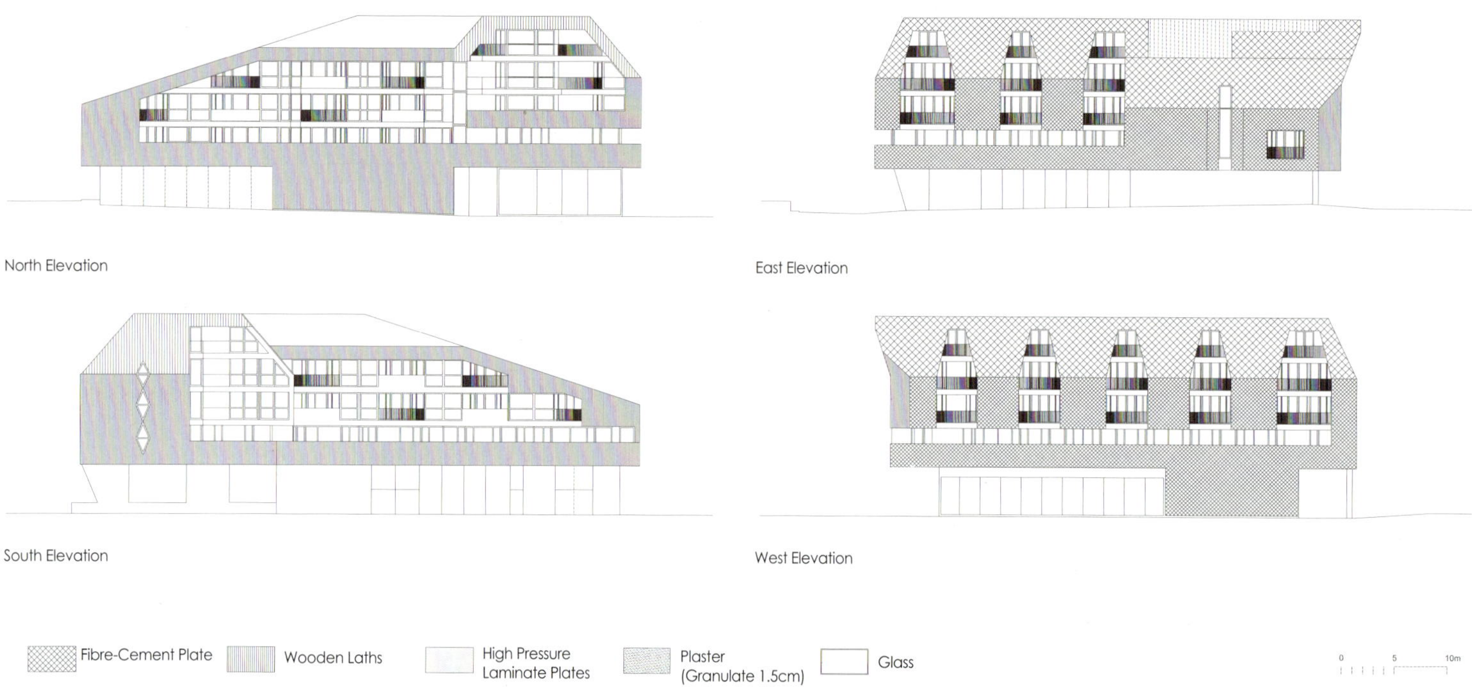

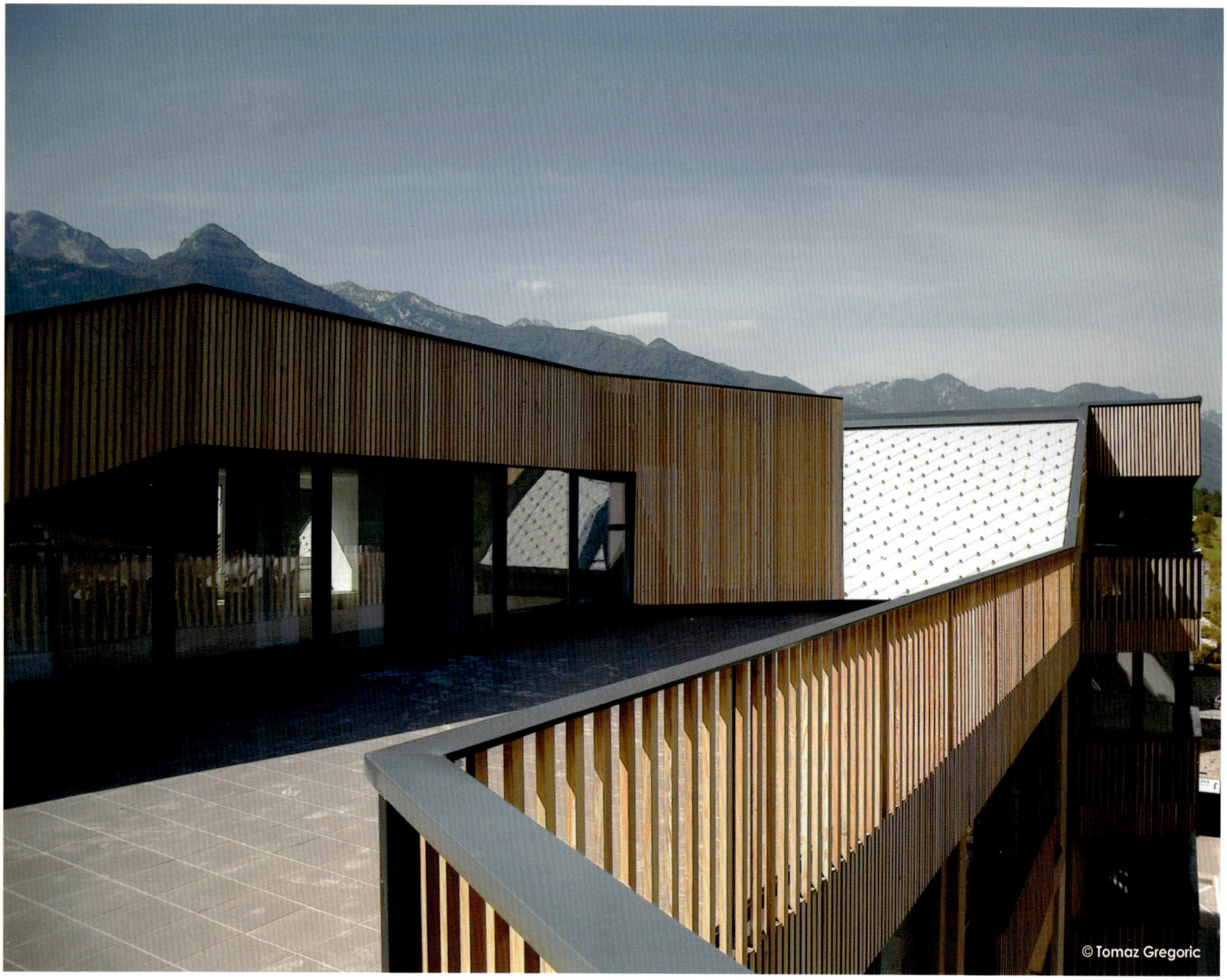
© Tomaz Gregoric

© Tomaz Gregoric

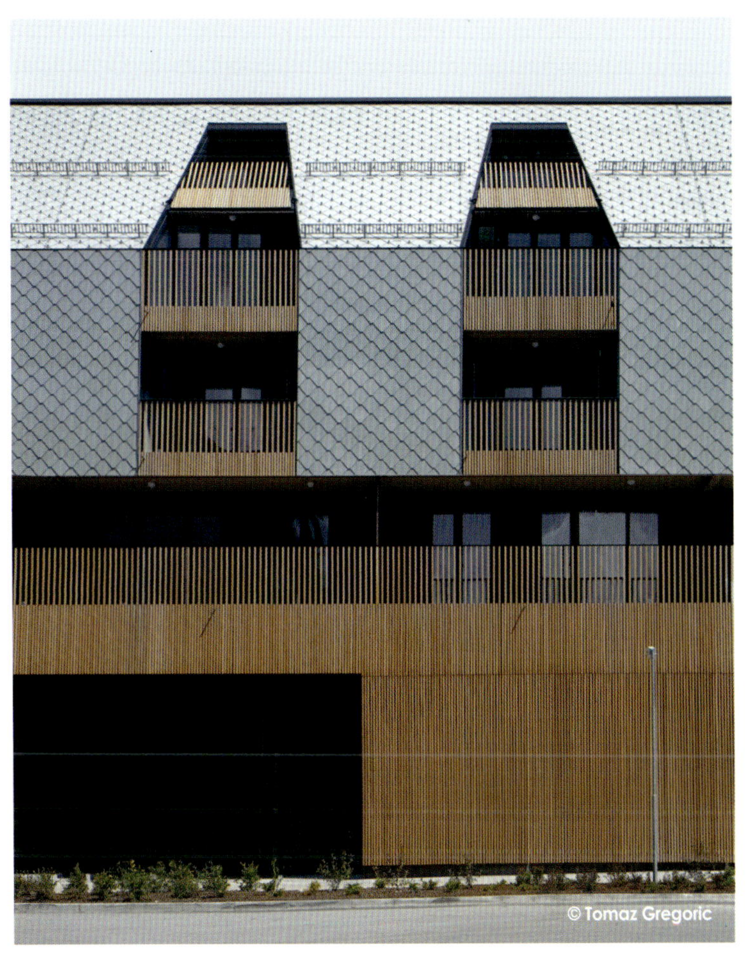

© Tomaz Gregoric

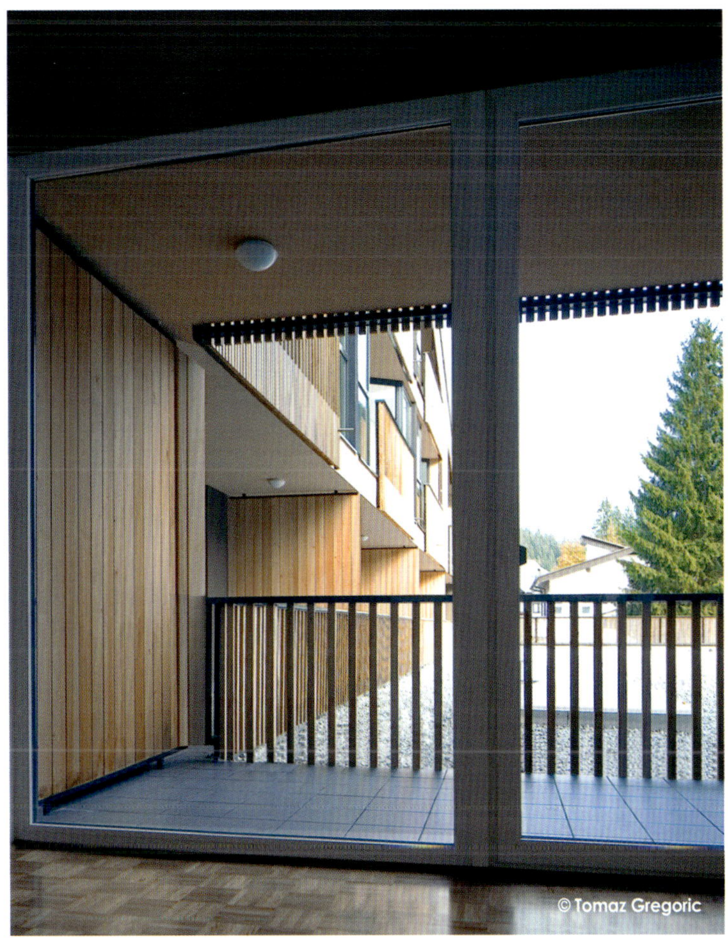

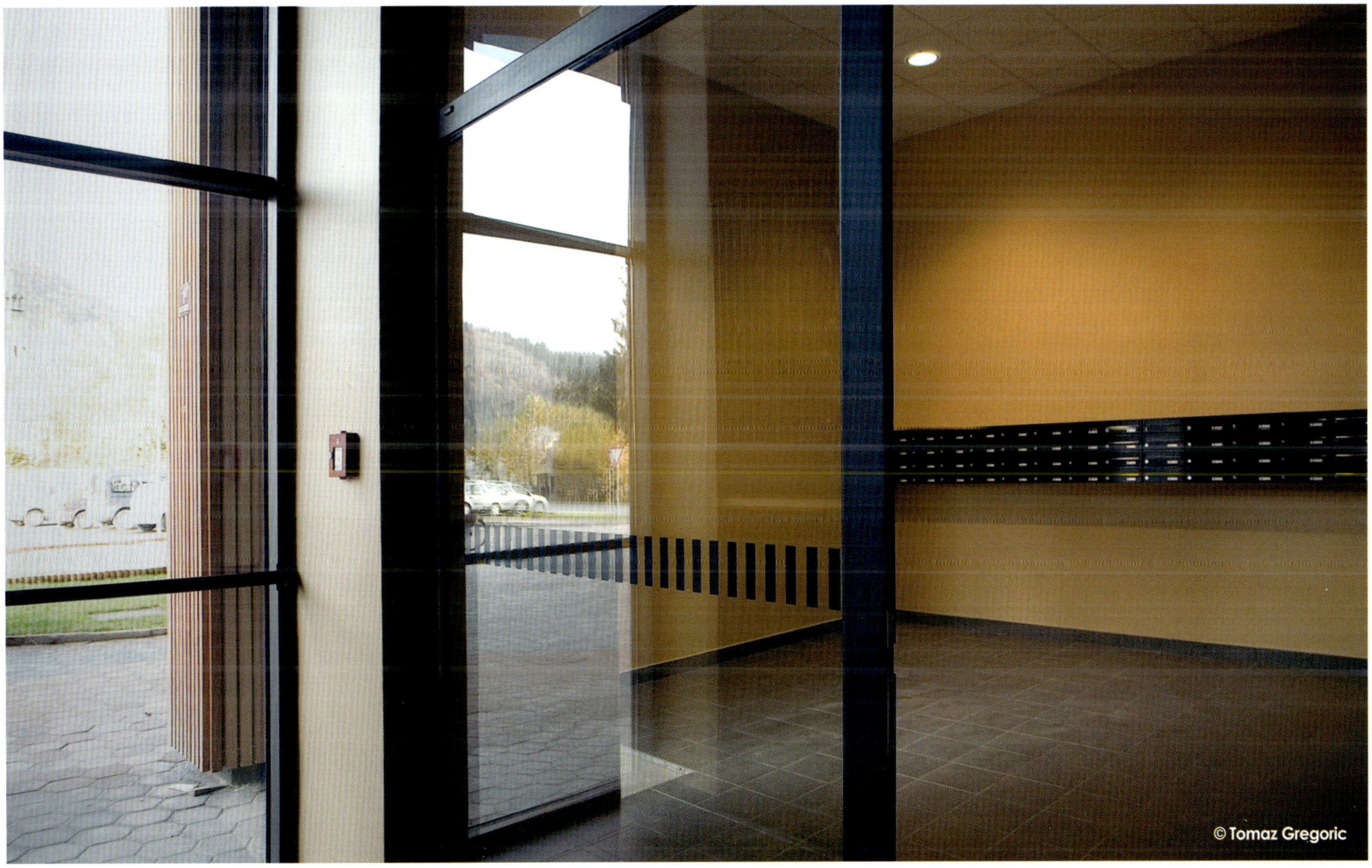
© Tomaz Gregoric

TETRIS APARTMENTS, Ljubljana, Slovenia

OFIS arhitekti

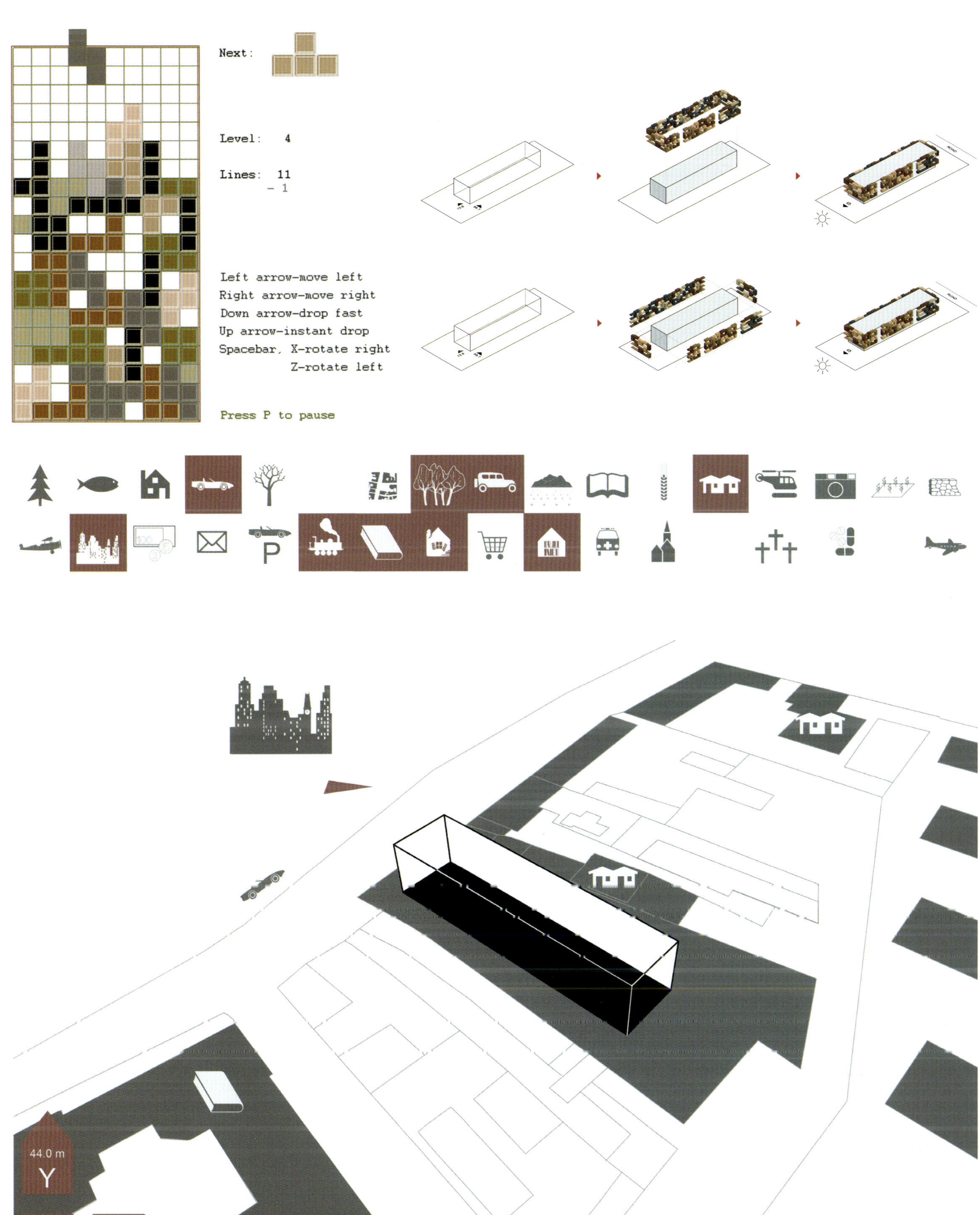
Next:
Level: 4
Lines: 11
- 1
Left arrow-move left
Right arrow-move right
Down arrow-drop fast
Up arrow-instant drop
Spacebar, X-rotate right
Z-rotate left
Press P to pause
44.0 m
Y
5 FLOOR
Z
53.0 m
X

© Tomaz Gregoric

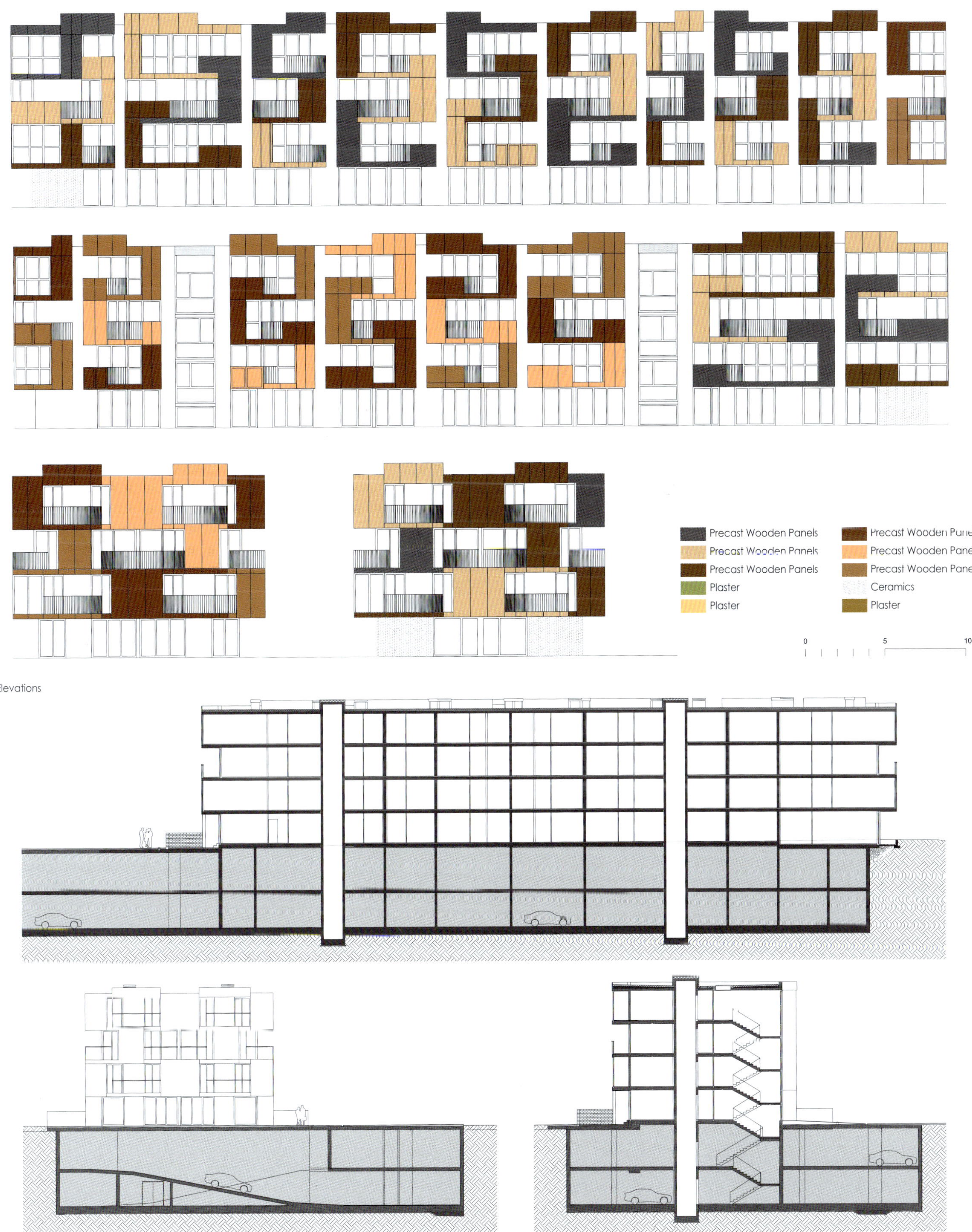

Elevations

Section

Aldeburgh Music Creative Campus, Suffolk, UK

Haworth Tompkins

© Philip Vile

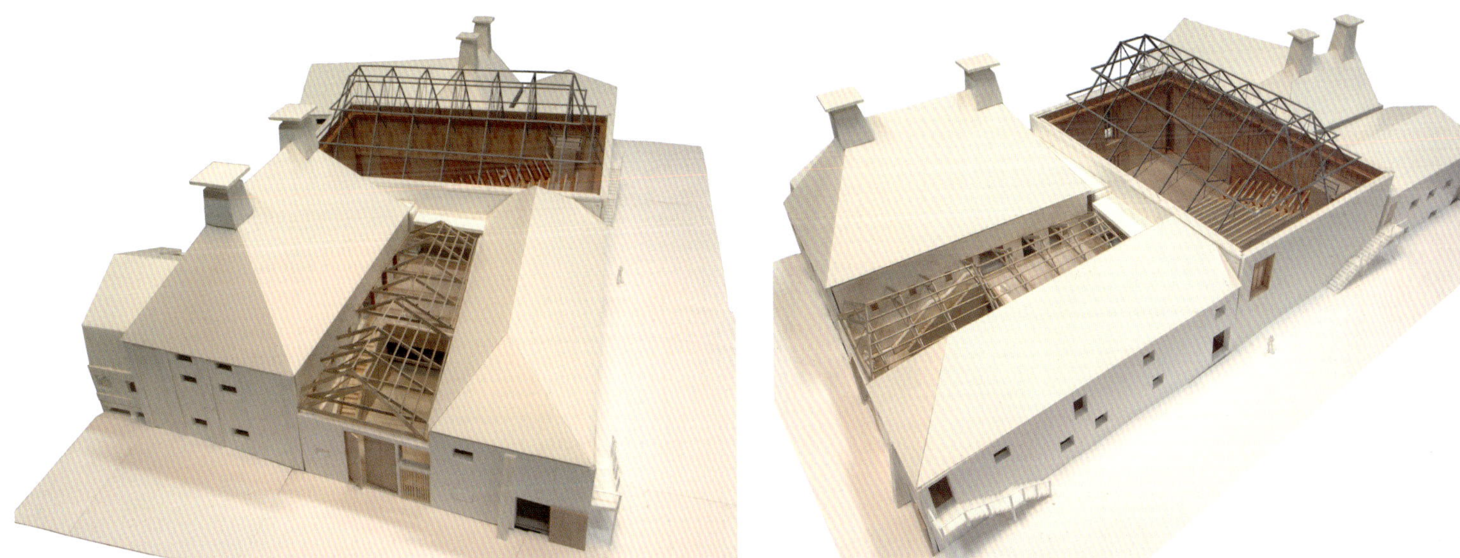

Study Modeling

West - East Section Looking North

Section

1. Entrance lobby
2. Office
3. WCs
4. Kitchen and servery area
5. Foyer
6. Multipurpose room
7. Back stage area
8. Britten Studio
9. Store
10. Plant
11. Mezzanine
12. Jerwood Kiln Studio
13. Winch room
14. Courtyard

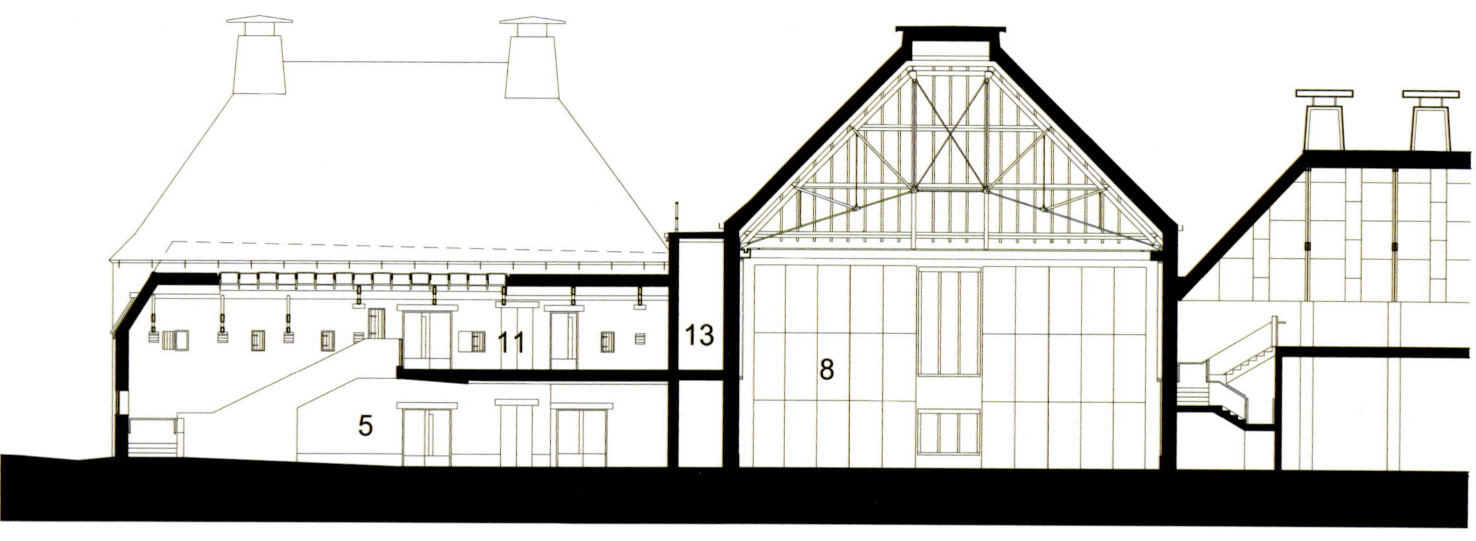

South - North Section Looking West

Construction Process

3.

4.

5.

6.

7.

Ceiling Construction Process

© Philip Vile

Clover House, Okazaki, Aichi, Japan

MAD Architects

© Fuji Koji

© Fuji Koji

© Fuji Koji

© Fuji Koji

Interior Diagram

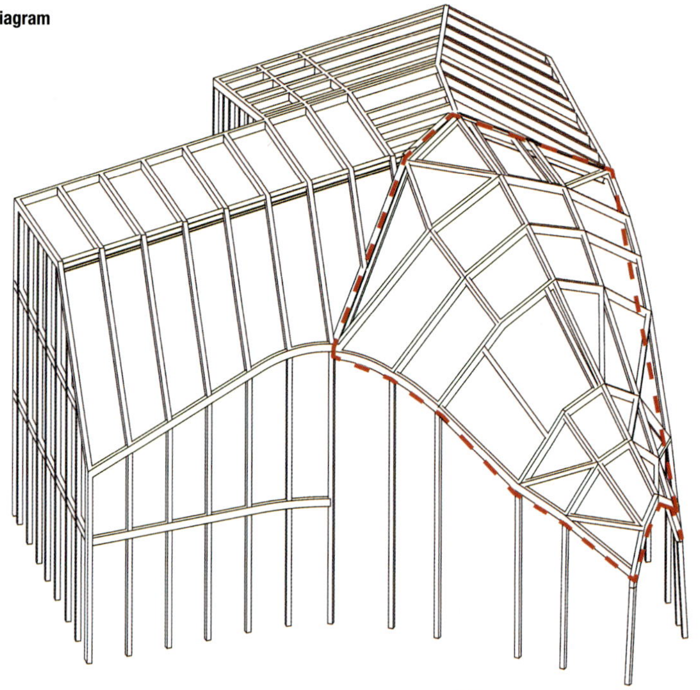

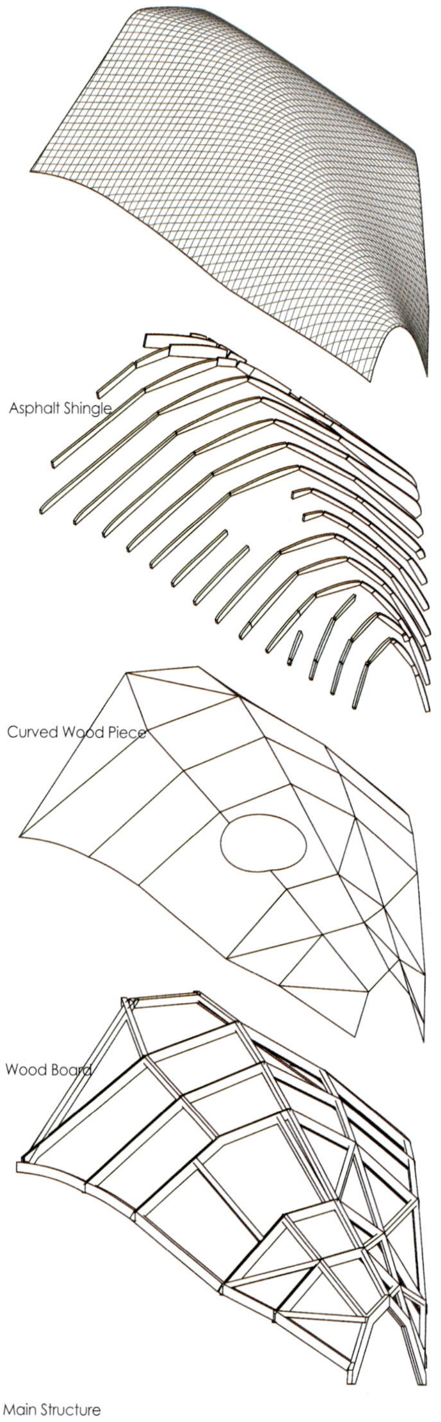

Asphalt Shingle

Curved Wood Piece

Wood Board

Main Structure

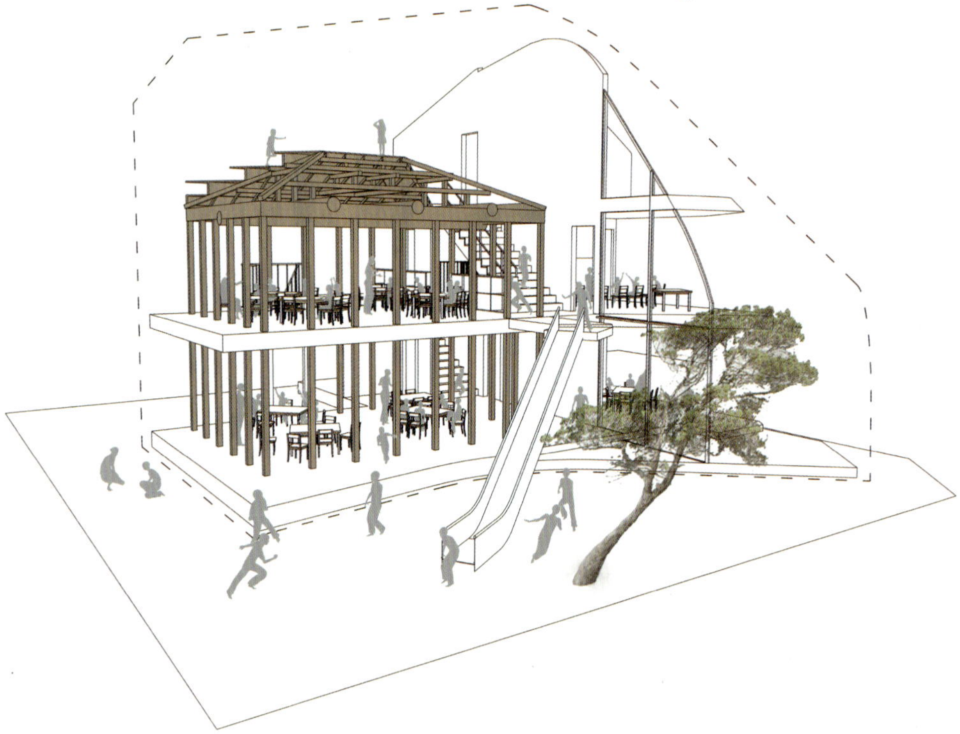

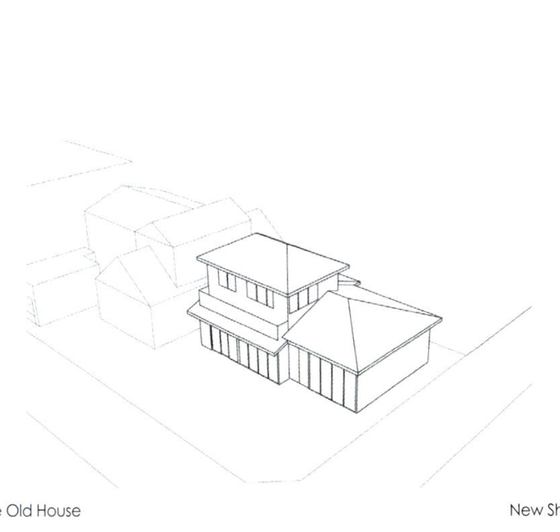

The Old House

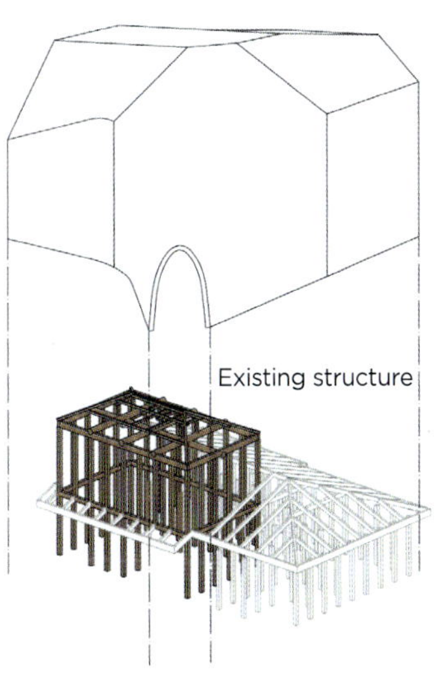

New Shell

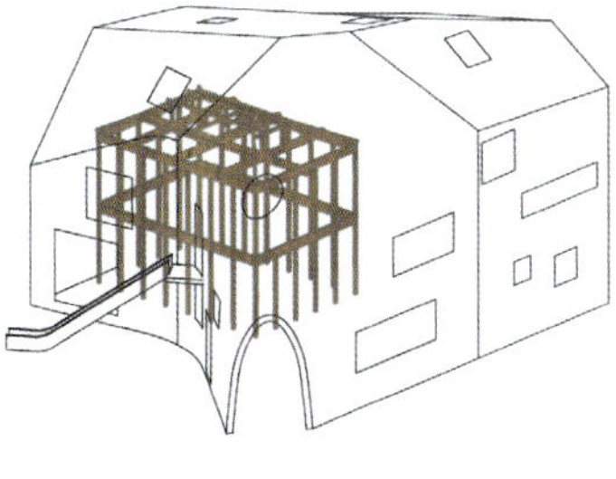

New Shell With Exosting Structure

Construction Process

© Fuji Koji

© Fuji Koji

© Fuji Koji

MARITIME YOUTH HOUSE, Copenhagen, Denmark

JDS Architects

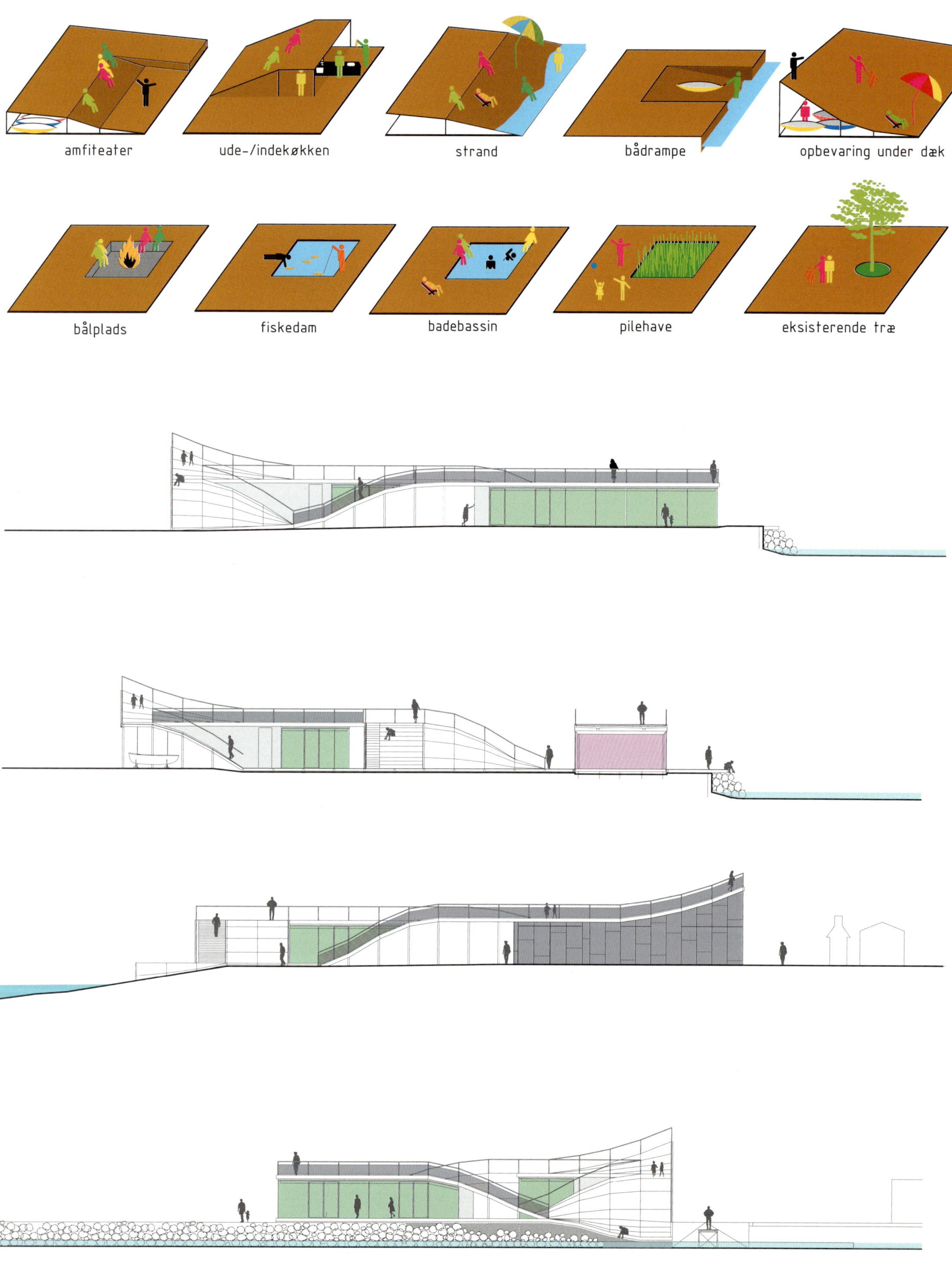
amfiteater
ude-/indekøkken
strand
bådrampe
opbevaring under dæk
bålplads
fiskedam
badebassin
pilehave
eksisterende træ

Elevation

© Paolo Rosselli

© Mads Hilmer

© Paolo Rosselli

© Paolo Rosselli

© Mads Hilmer

b_shop, Seoul, Korea
Ehwa Yoo

© Jeong Tae-ho

© Jeong Tae-ho

© Jeong Tae-ho

Binet Business Center, Paris, France
AZC

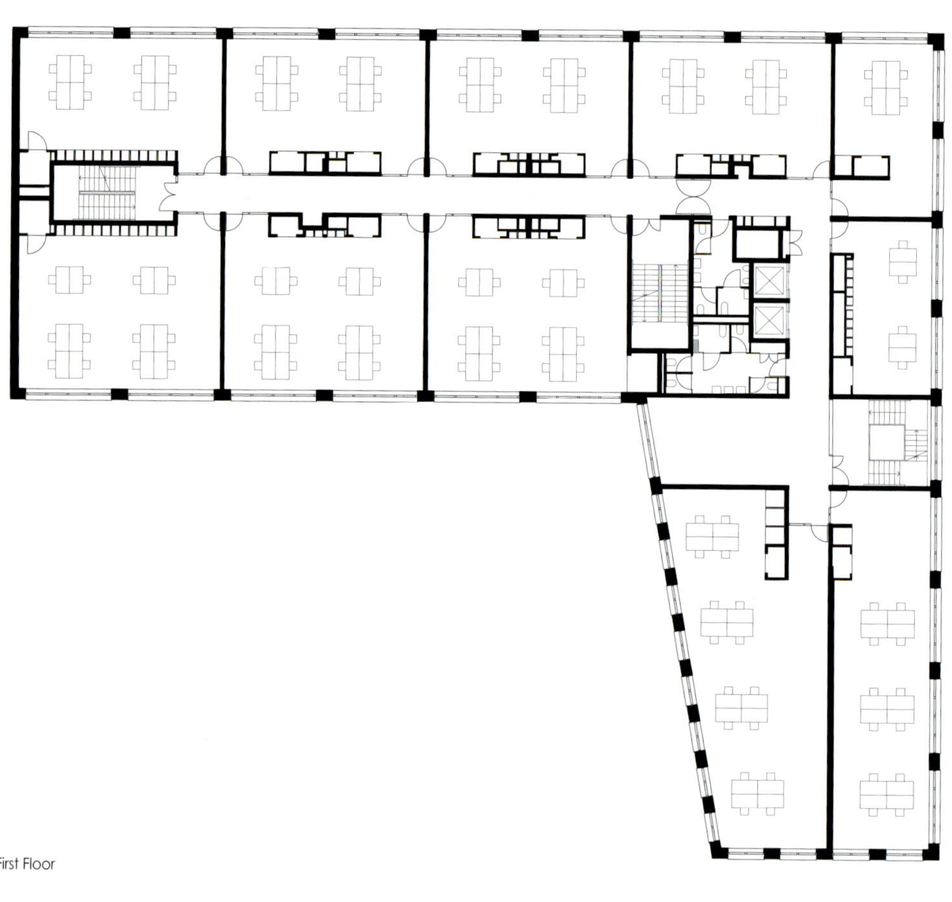

First Floor

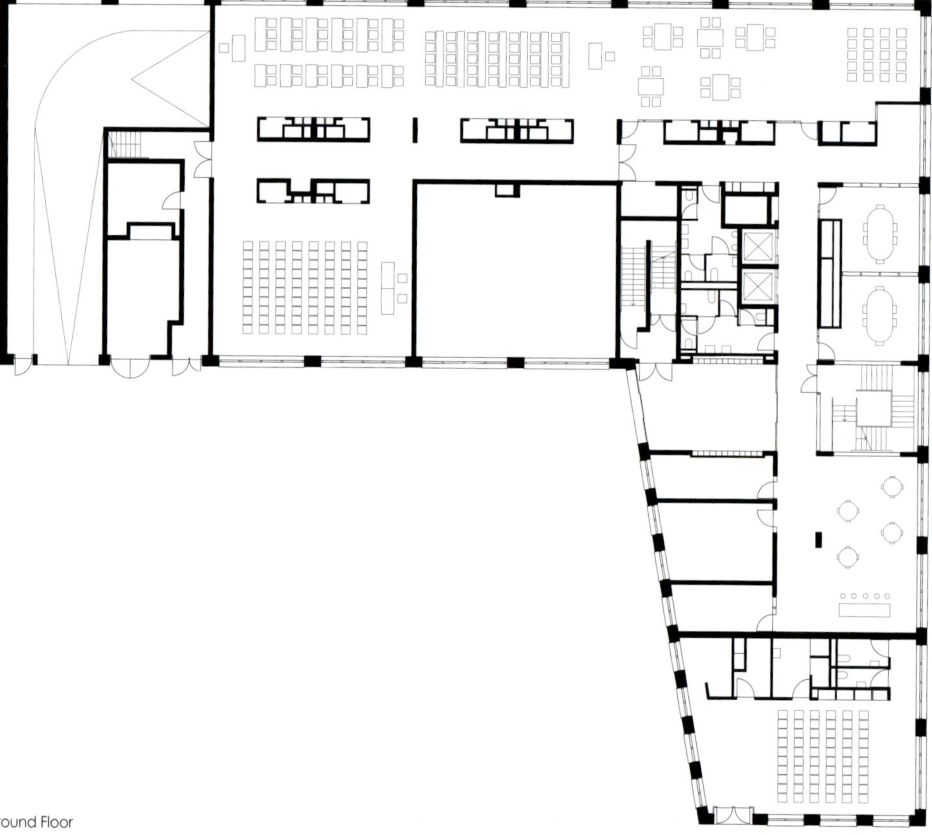

Ground Floor

Box in the Box, Madrid, Spain
Arenas Basabe Palacios Arquitectos

© Imagen Subliminal

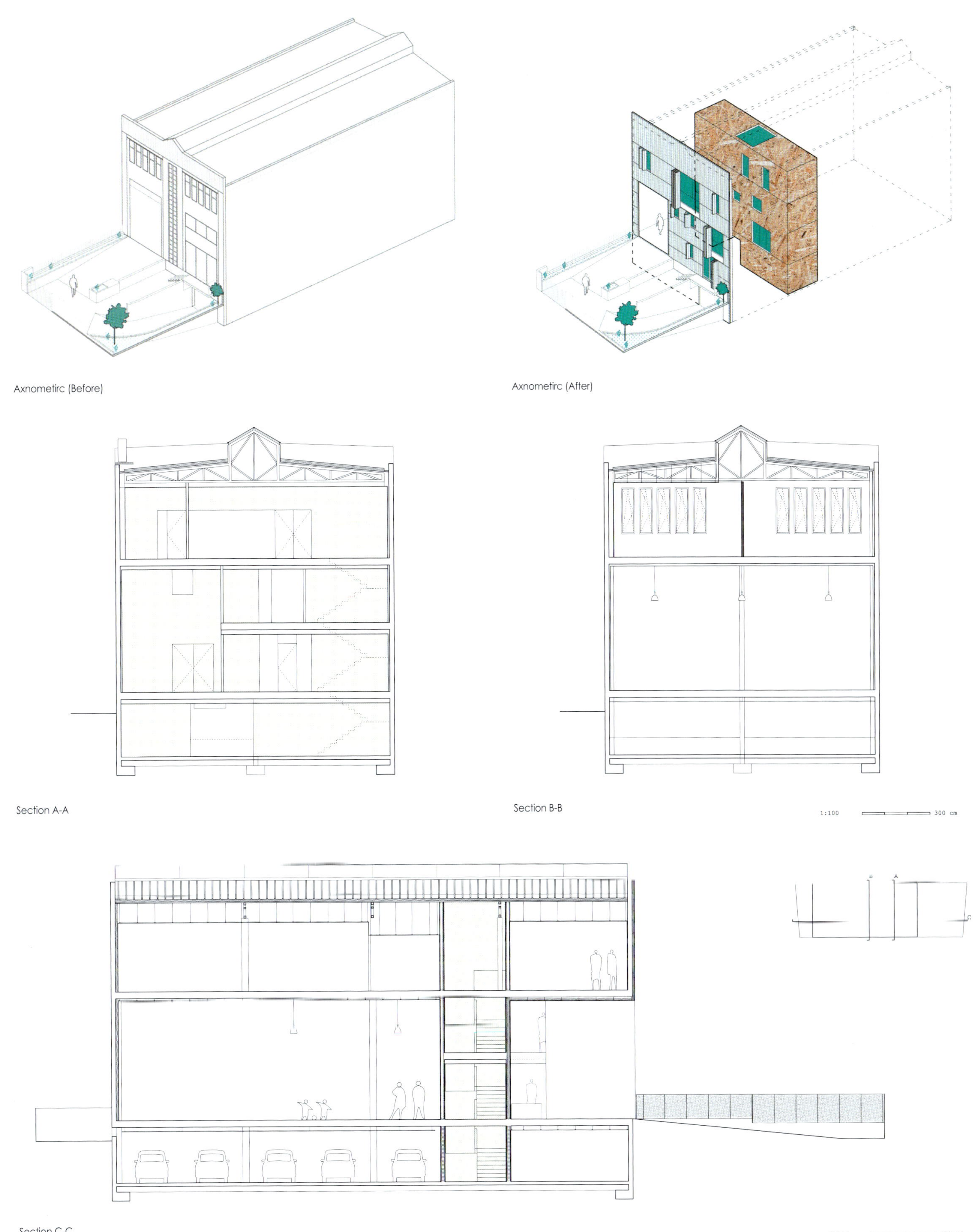
Axnometirc (Before)
Axnometirc (After)
Section A-A
Section B-B
1:100
300 cm
Section C-C
1:100
300 cm

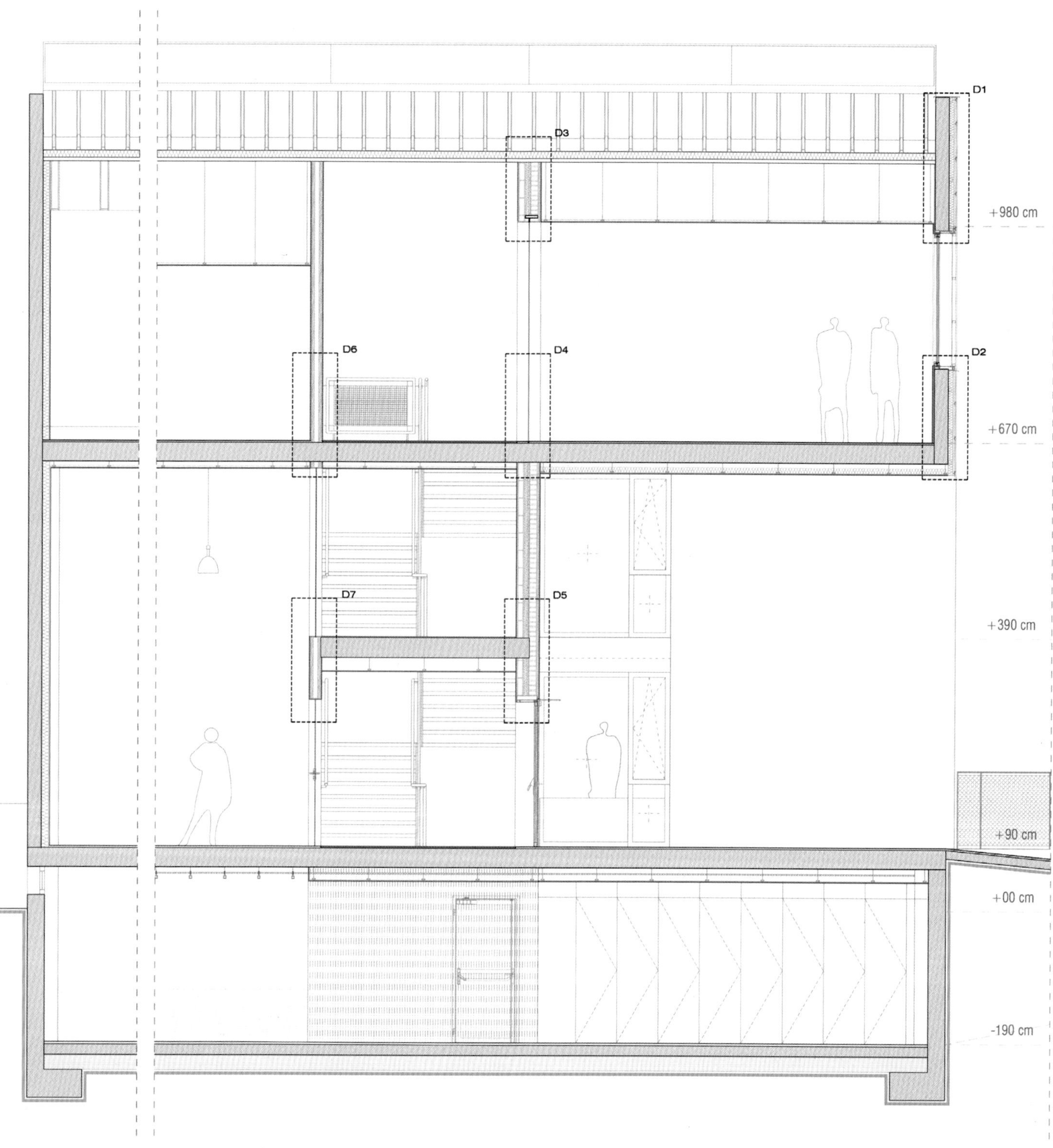

Section 1:50

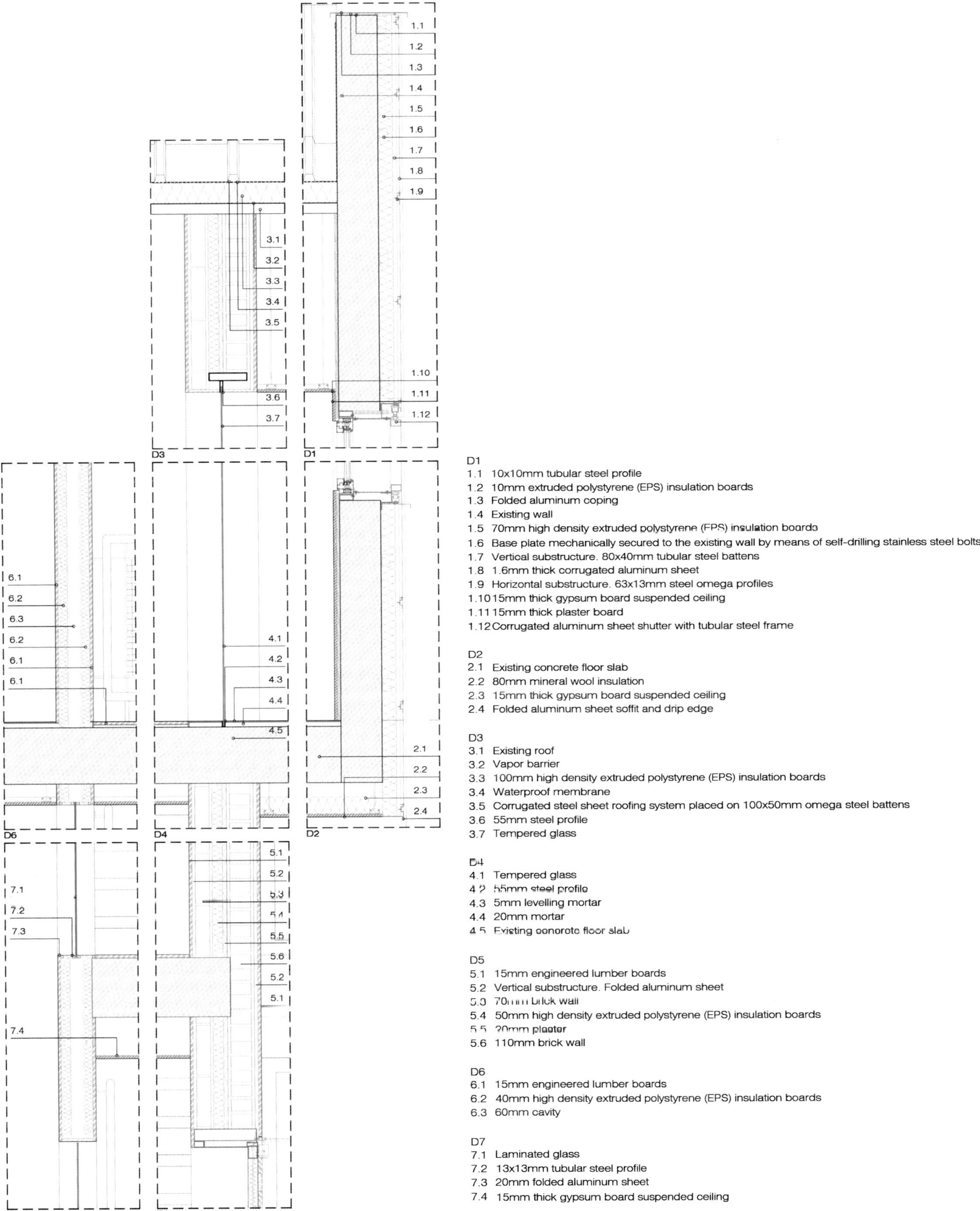

D1
1.1 10x10mm tubular steel profile
1.2 10mm extruded polystyrene (EPS) insulation boards
1.3 Folded aluminum coping
1.4 Existing wall
1.5 70mm high density extruded polystyrene (EPS) insulation boards
1.6 Base plate mechanically secured to the existing wall by means of self-drilling stainless steel bolts
1.7 Vertical substructure. 80x40mm tubular steel battens
1.8 1.6mm thick corrugated aluminum sheet
1.9 Horizontal substructure. 63x13mm steel omega profiles
1.10 15mm thick gypsum board suspended ceiling
1.11 15mm thick plaster board
1.12 Corrugated aluminum sheet shutter with tubular steel frame

D2
2.1 Existing concrete floor slab
2.2 80mm mineral wool insulation
2.3 15mm thick gypsum board suspended ceiling
2.4 Folded aluminum sheet soffit and drip edge

D3
3.1 Existing roof
3.2 Vapor barrier
3.3 100mm high density extruded polystyrene (EPS) insulation boards
3.4 Waterproof membrane
3.5 Corrugated steel sheet roofing system placed on 100x50mm omega steel battens
3.6 55mm steel profile
3.7 Tempered glass

D4
4.1 Tempered glass
4.2 55mm steel profile
4.3 5mm levelling mortar
4.4 20mm mortar
4.5 Existing concrete floor slab

D5
5.1 15mm engineered lumber boards
5.2 Vertical substructure. Folded aluminum sheet
5.3 70mm brick wall
5.4 50mm high density extruded polystyrene (EPS) insulation boards
5.5 20mm plaster
5.6 110mm brick wall

D6
6.1 15mm engineered lumber boards
6.2 40mm high density extruded polystyrene (EPS) insulation boards
6.3 60mm cavity

D7
7.1 Laminated glass
7.2 13x13mm tubular steel profile
7.3 20mm folded aluminum sheet
7.4 15mm thick gypsum board suspended ceiling

Section Detail 1:15

© Imagen Subliminal

© Imagen Subliminal

Clemente Dental Clinic, Madrid, Spain

LANDÍNEZ+REY arquitectos

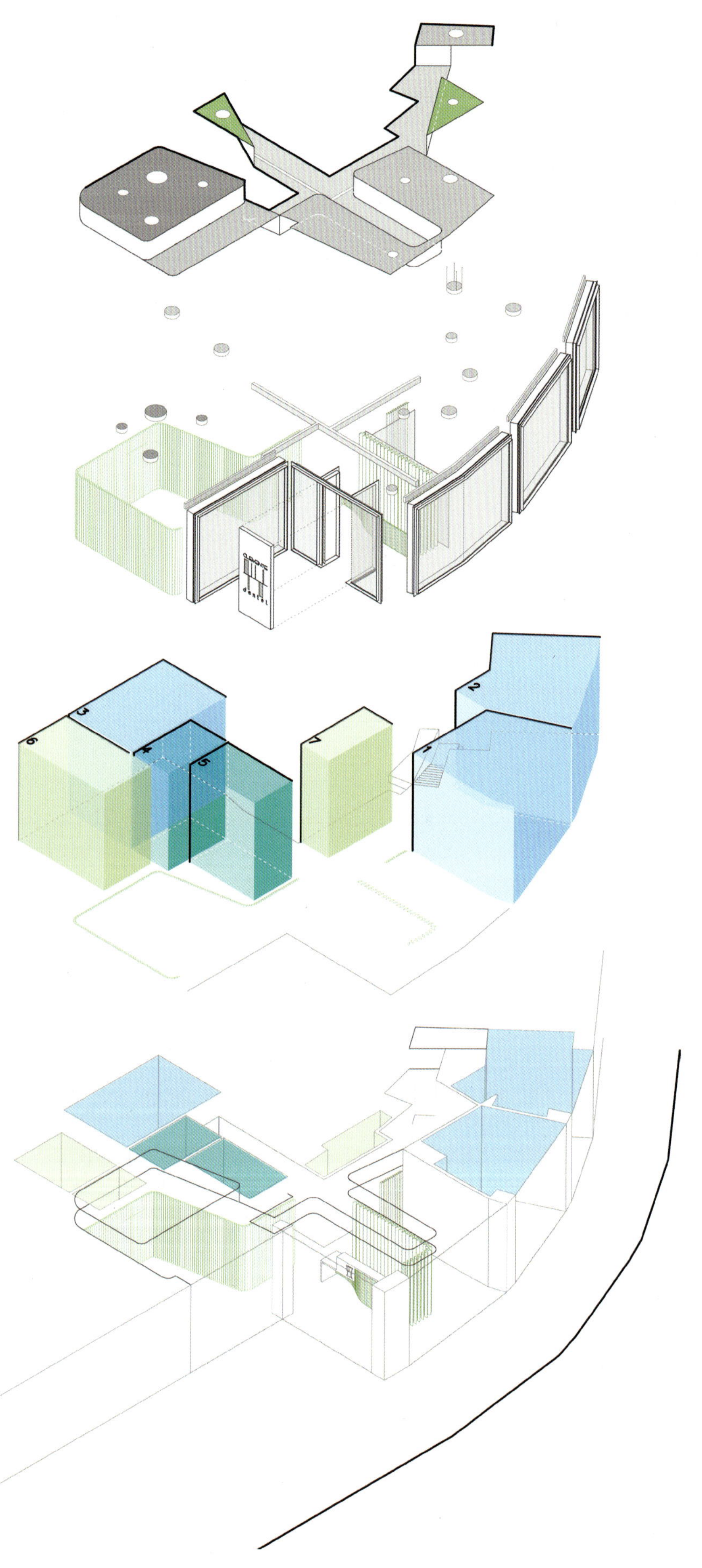

GABINETE ODONTOLÓGICO | 1 2 3

SALA ESTERELIZACIÓN | 4

SALA RAYOS X PANORÁMICO | 5

ASEOS PÚBLICOS | 6

APOYO | 7

Diagram

© gustavo GONZÁLEZ BELLÓN

© gustavo GONZÁLEZ BELLÓN

© gustavo GONZÁLEZ BELLÓN

La Dynamo de Banlieues Bleues, Pantin, France

PERIPHERIQUES

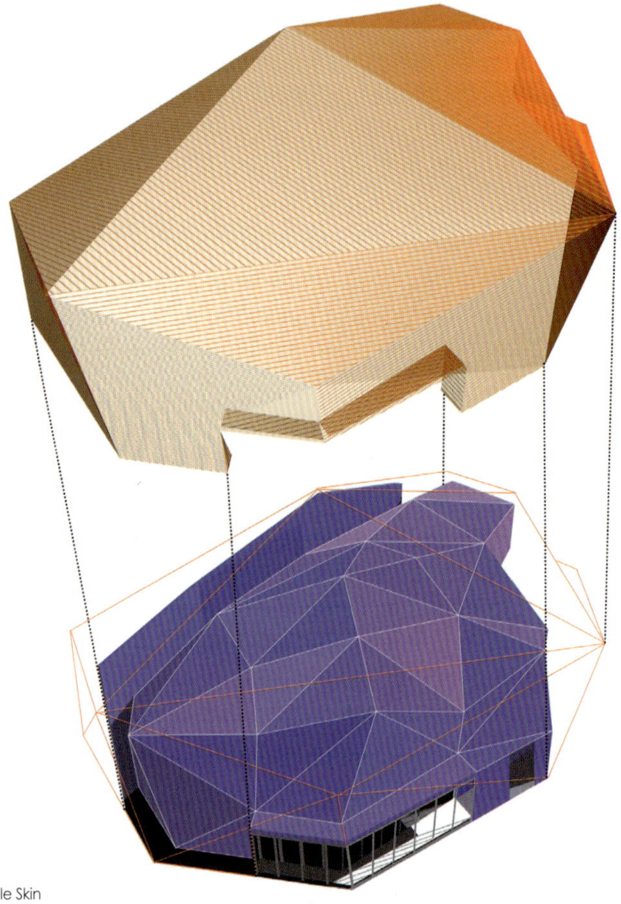

Double Skin

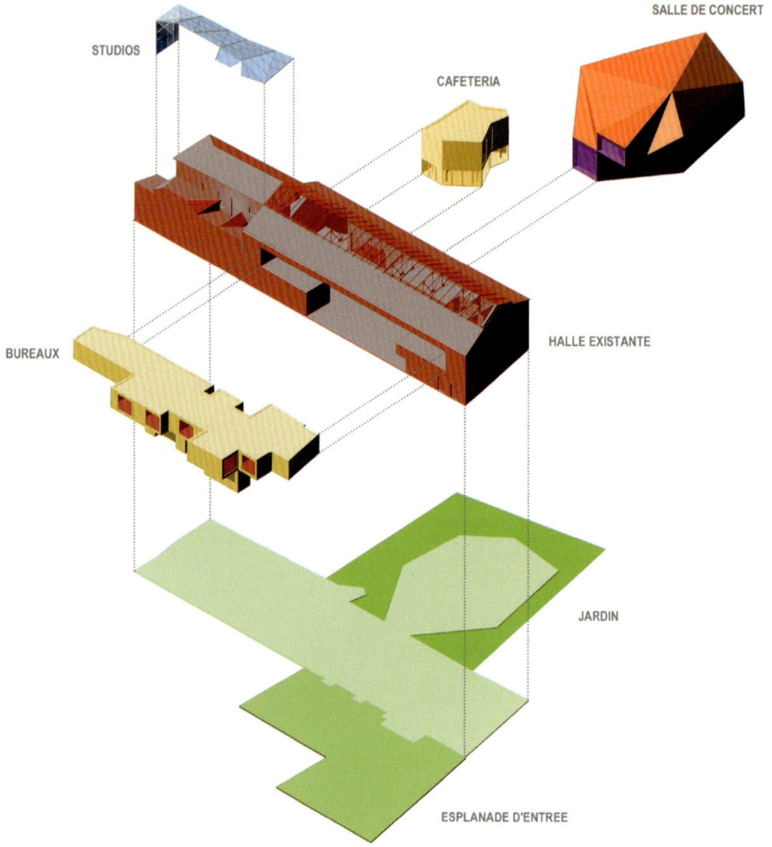

STUDIOS
CAFETERIA
SALLE DE CONCERT
HALLE EXISTANTE
BUREAUX
JARDIN
ESPLANADE D'ENTREE

Axonometric

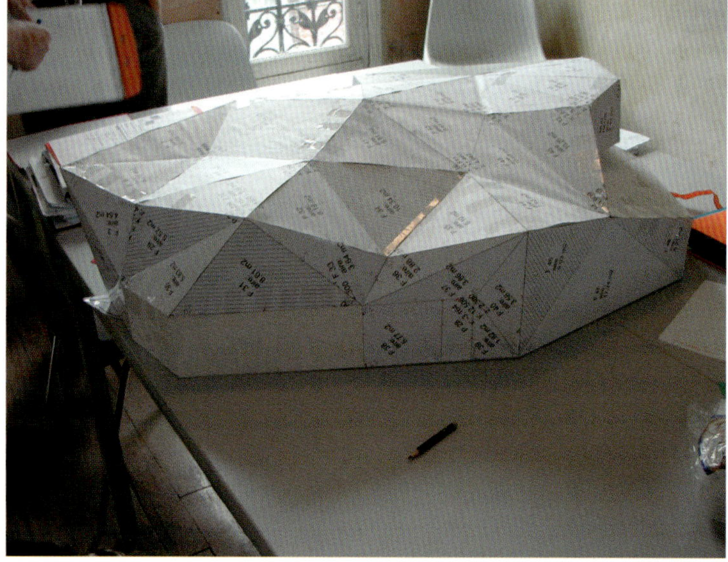

Study Modeling

1.

2.

3.

4.

Construction Process

Museo del Agua in Palencia, Palencia, Spain

MID estudio

© Helena Velez Olabarria

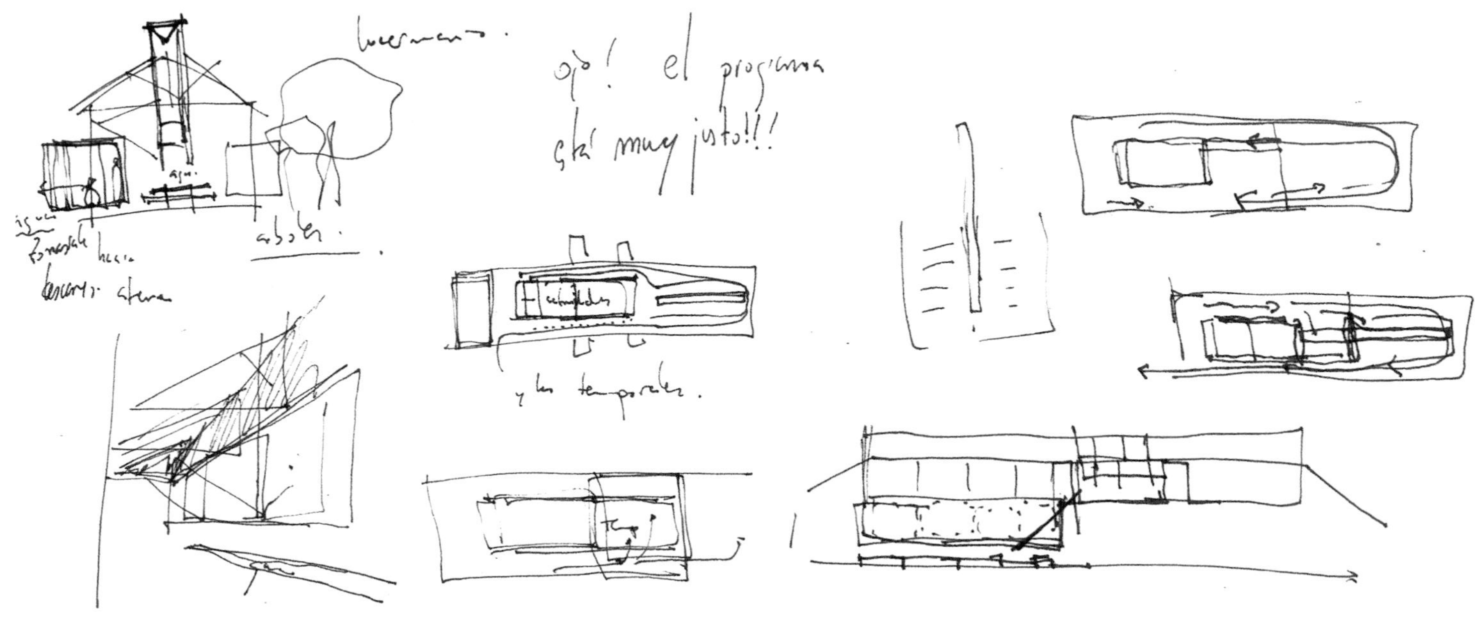

Idea Sketch

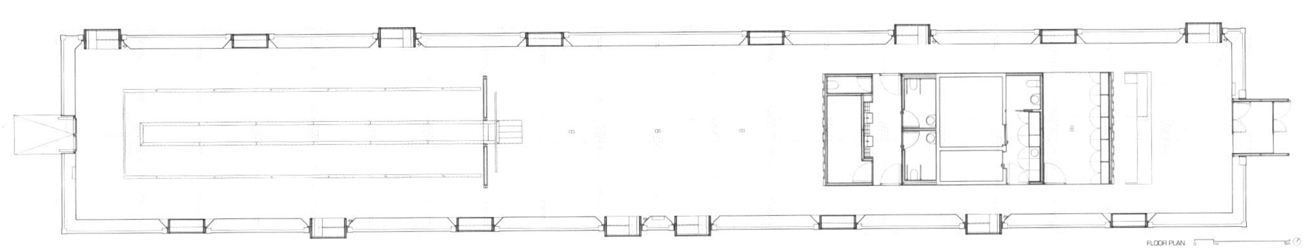

Ground Floor

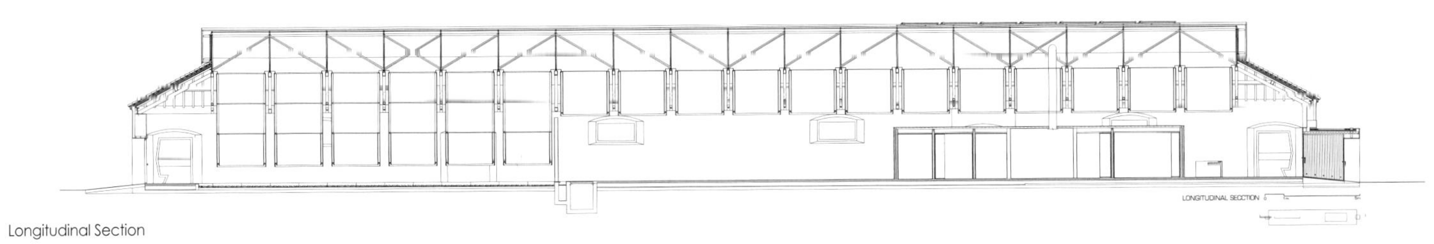

Longitudinal Section

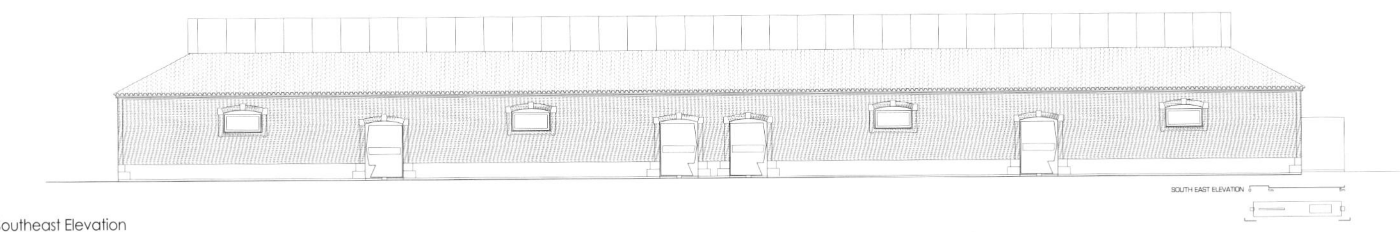

Southeast Elevation

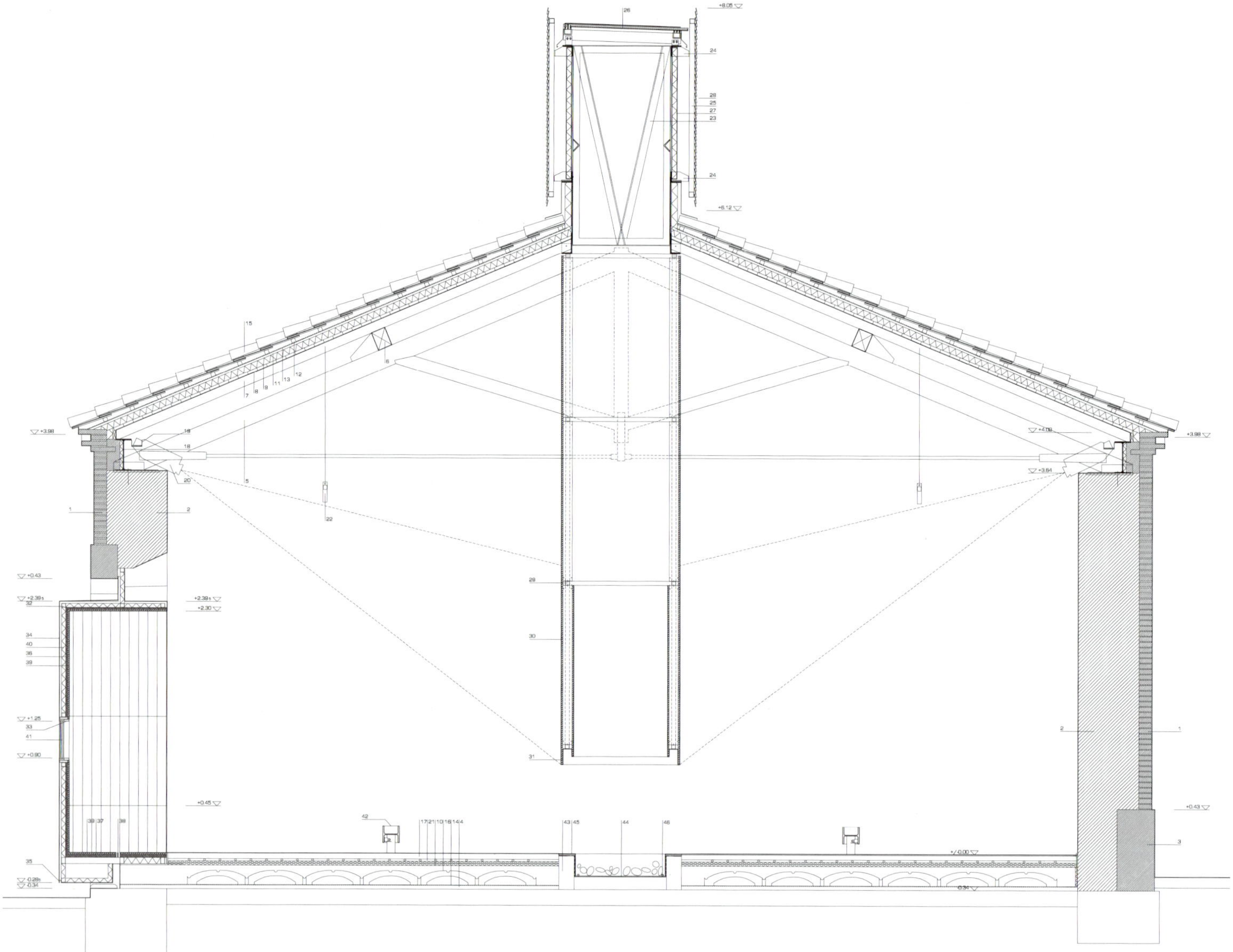

DETAIL CROSS SECTION 3

Idea Sketch

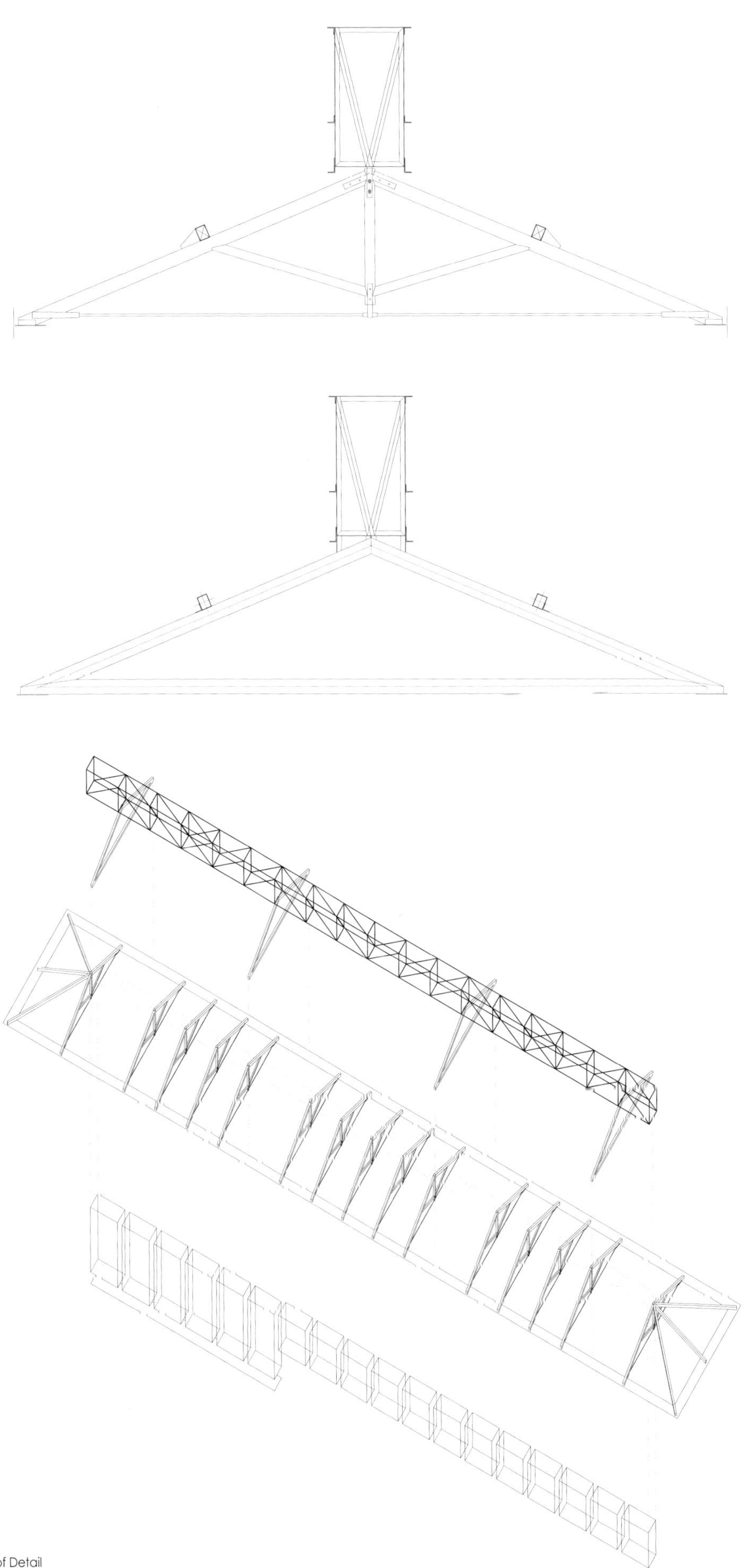

Roof Detail

1.

3.

2.

4.

Construction Process

© Helena Velez Olabarria

© Helena Velez Olabarria

Sustainable Club House, Lystrup, Denmark

CEBRA

© Mikkel Frost | CEBRA

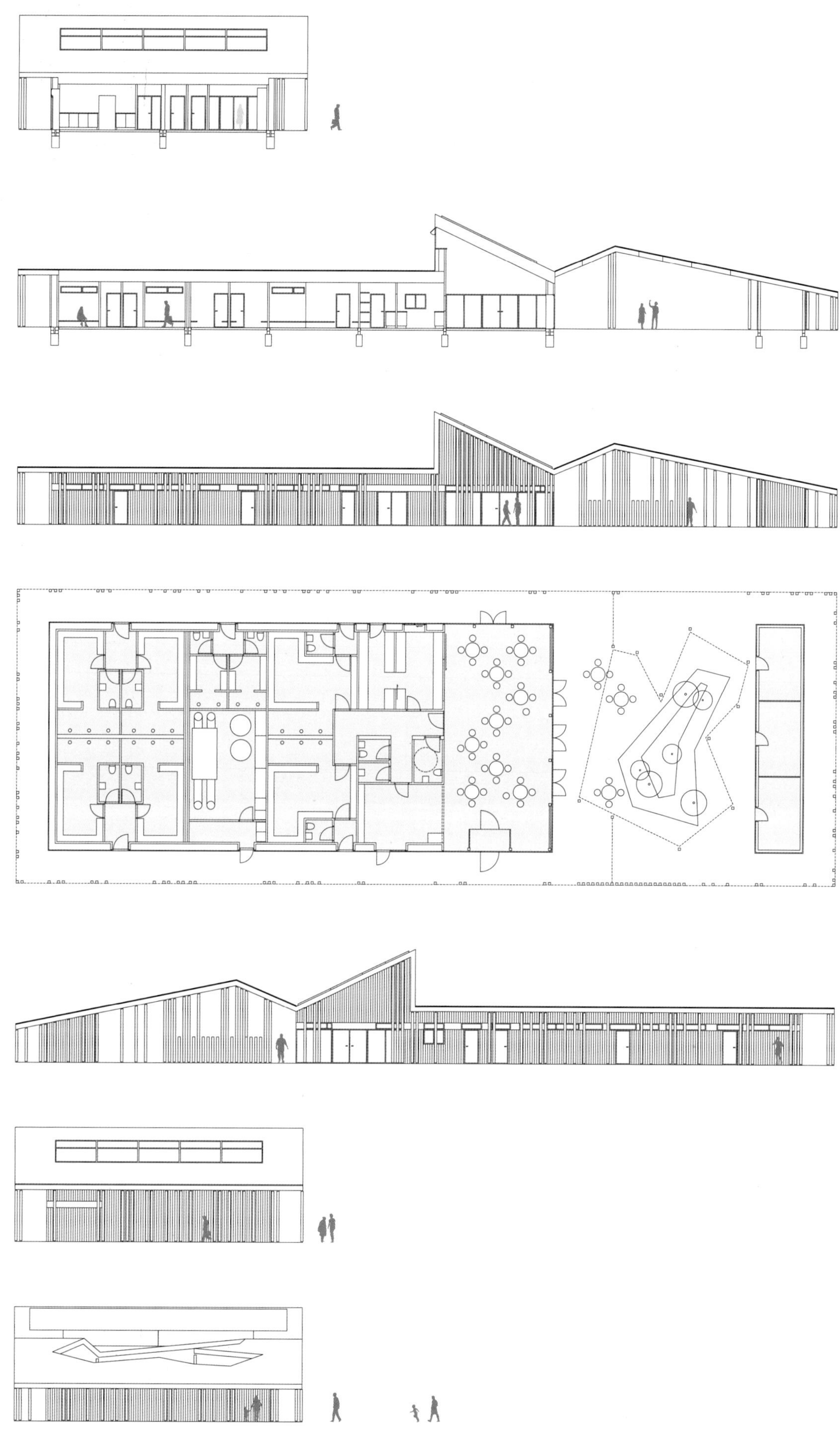

Elevation & Section

© Mikkel Frost | CEBRA

© Mikkel Frost | CEBRA

© Mikkel Frost | CEBRA

© Mikkel Frost | CEBRA

©Mikkel Frost | CEBRA

THE GOLFER, Isernia, Italy

Medir architects

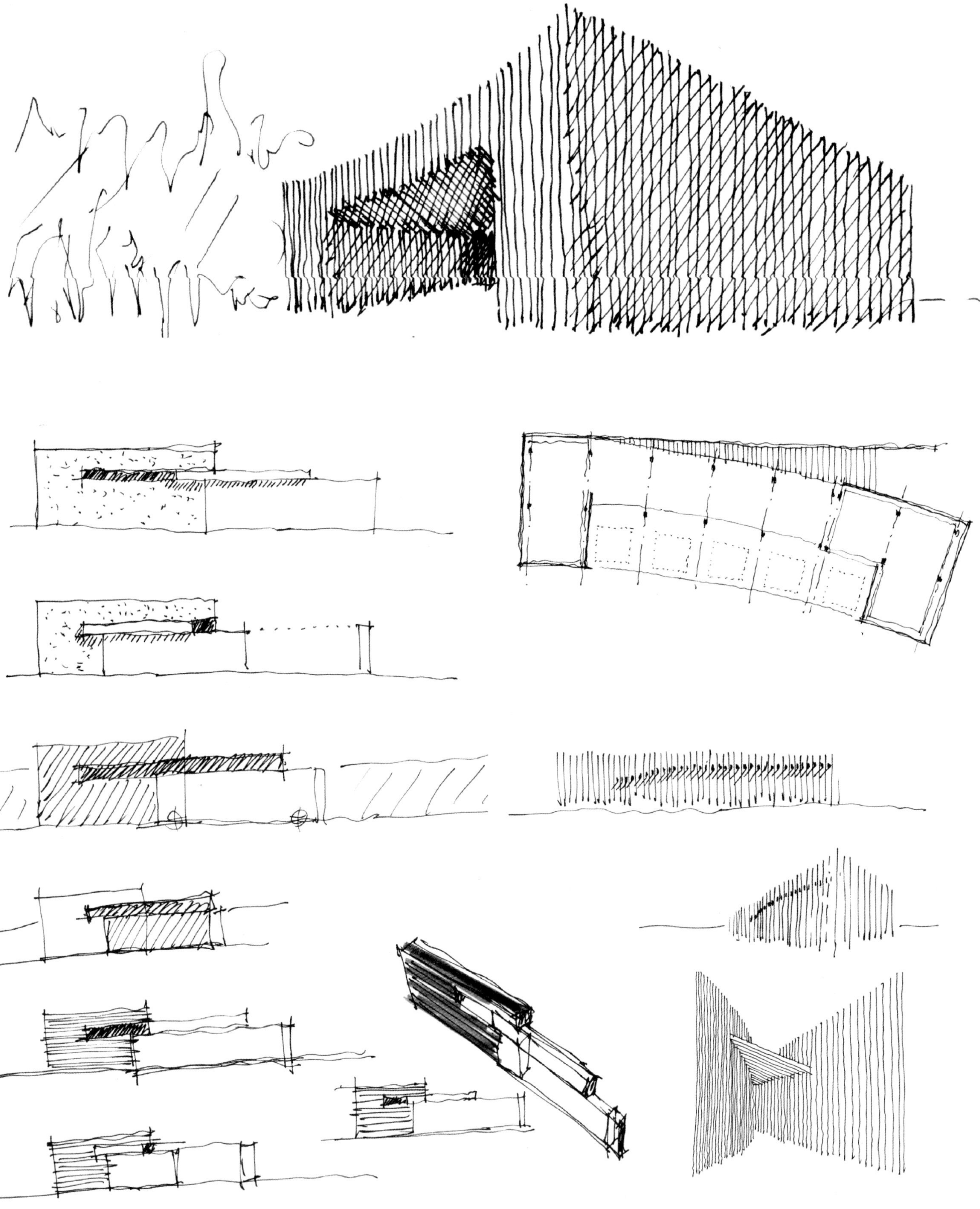
Idea Sketch

Uhl Foundation, Toscana, Italy

modostudio

Idea Sketch

General Plan

Study Modeling

© Laura Egger

© Laura Egger

© Laura Egger

21st june
9.00 AM
12.00 PM
18.00 PM
21st december
9.00 AM
12.00 PM
16.00 PM

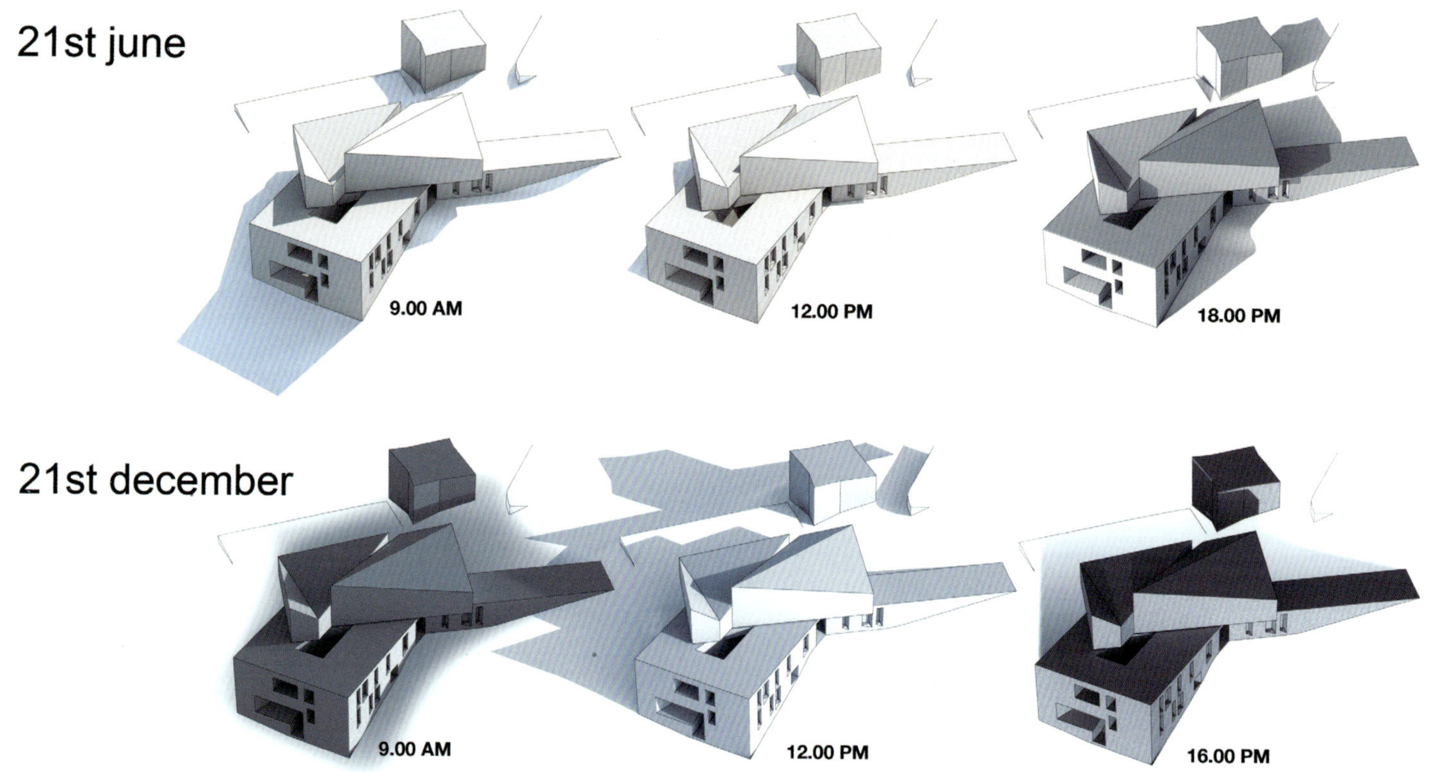

Sun Schemes

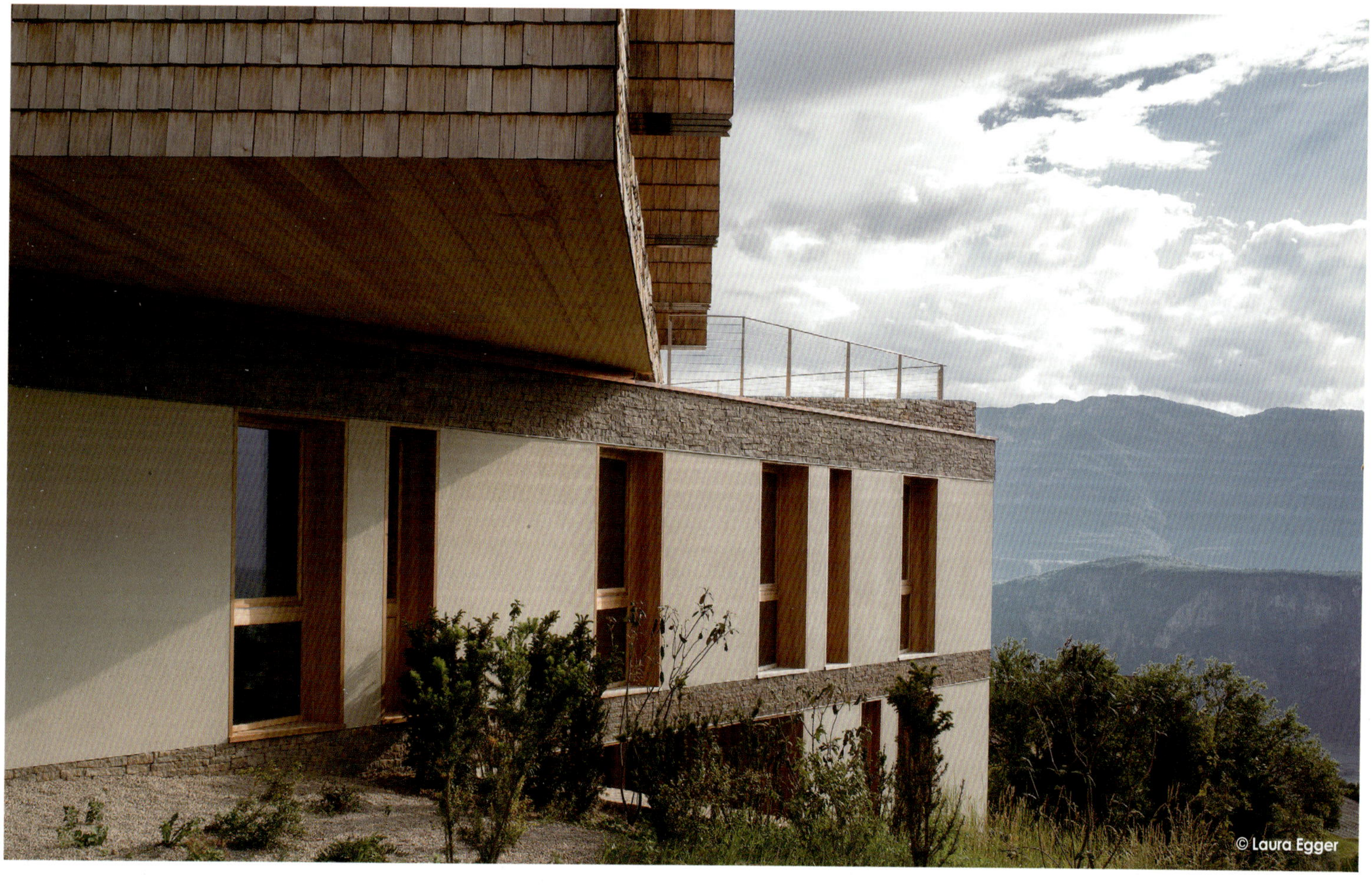

© Laura Egger

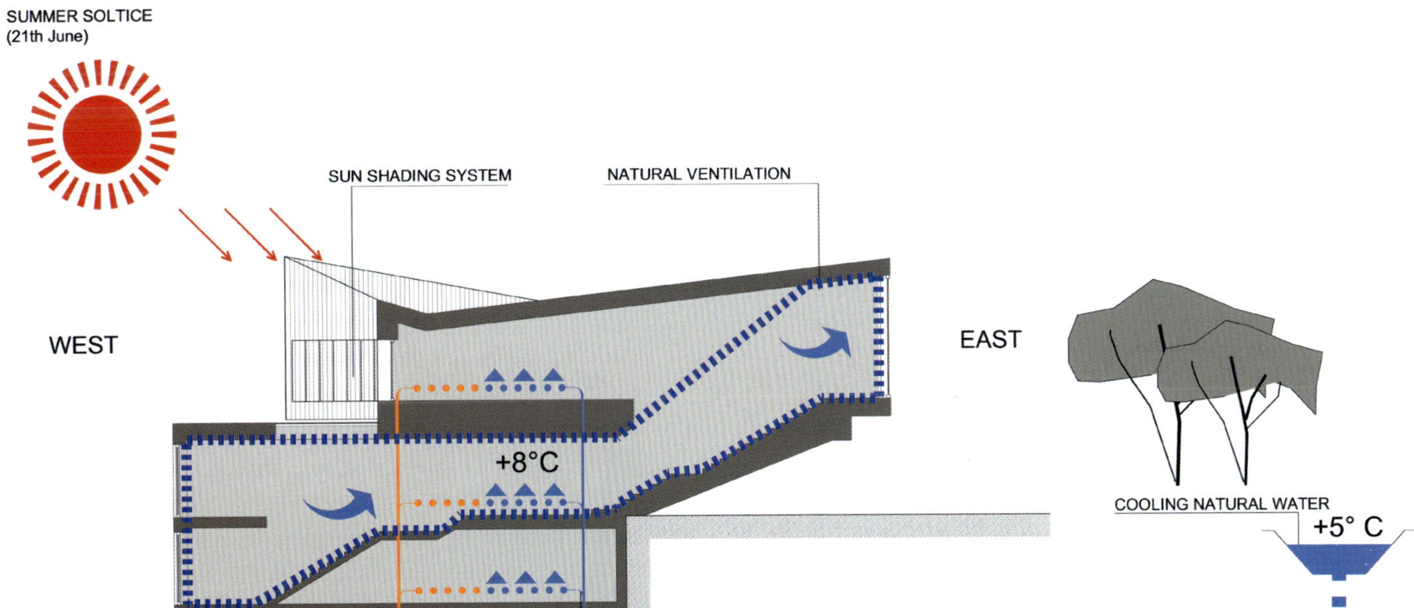

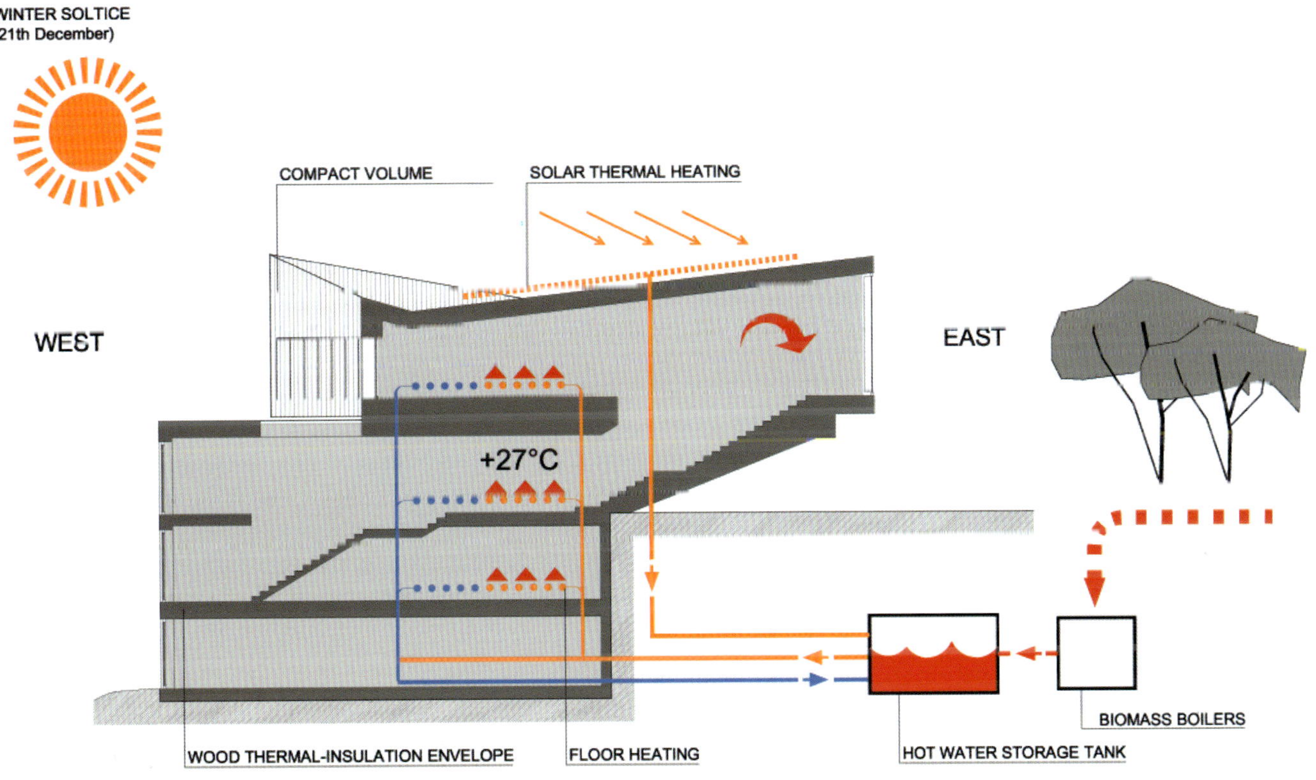

Energy Schemes

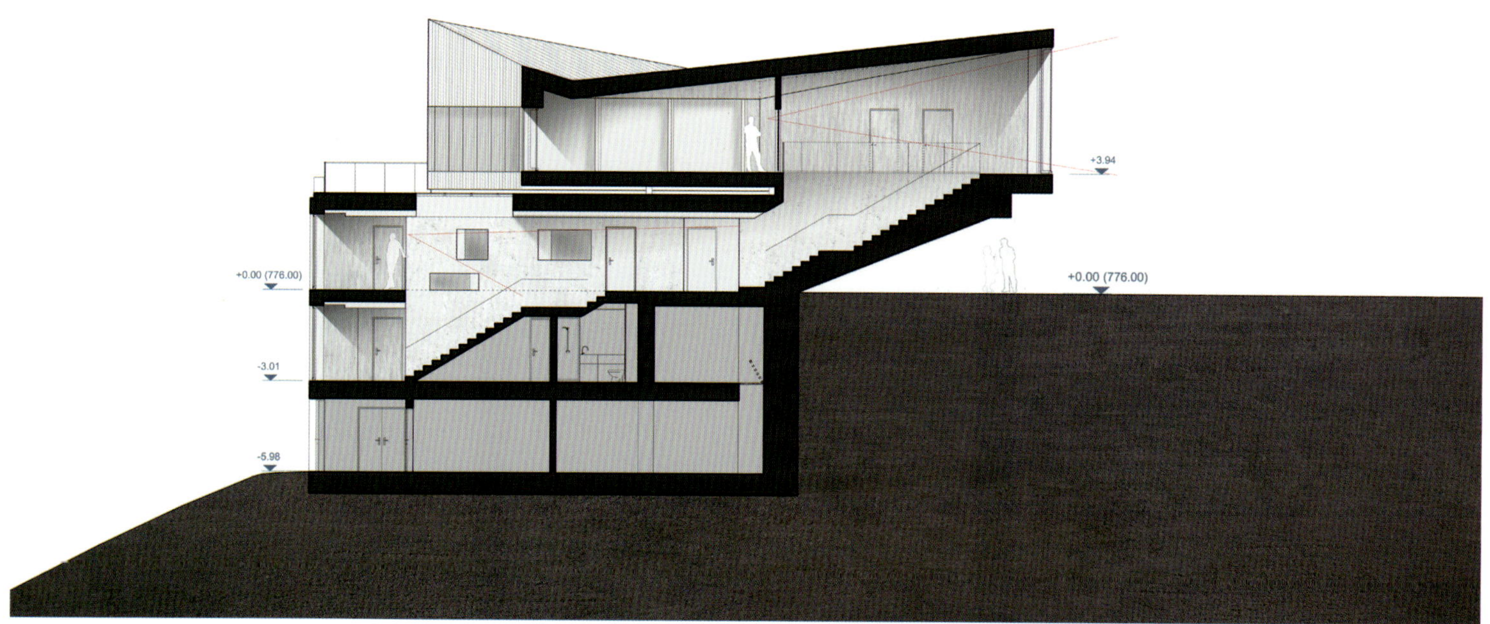

Section A

Section B

1.

2.

3.

4.

5.

6.

7.

8.

9.

Construction Process

© Laura Egger

© Laura Egger

© Laura Egger

© Laura Egger

Toigetation, Cao Bang Province, Vietnam
H&P Architects

© Doan Thanh Ha

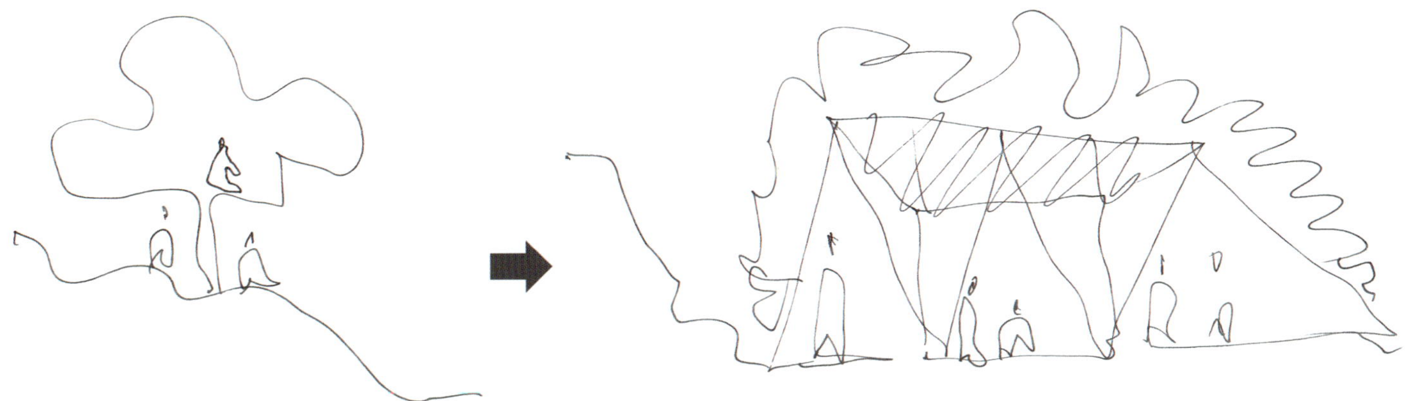

A large tree with wide spread canopy affording shade for the spaces below and within

Idea Diagram

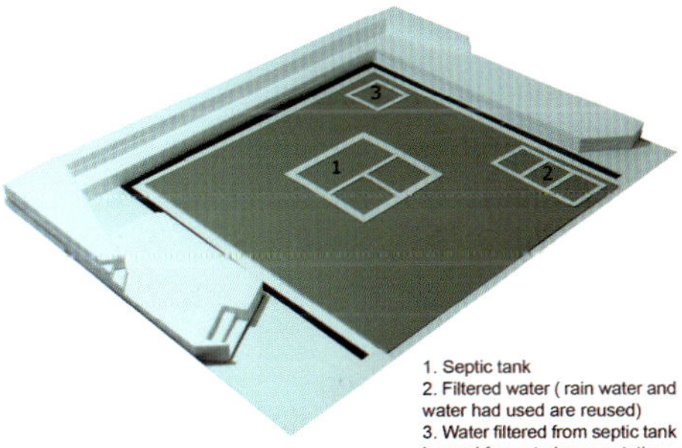

1. Septic tank
2. Filtered water (rain water and water had used are reused)
3. Water filtered from septic tank is used for watering vegetation

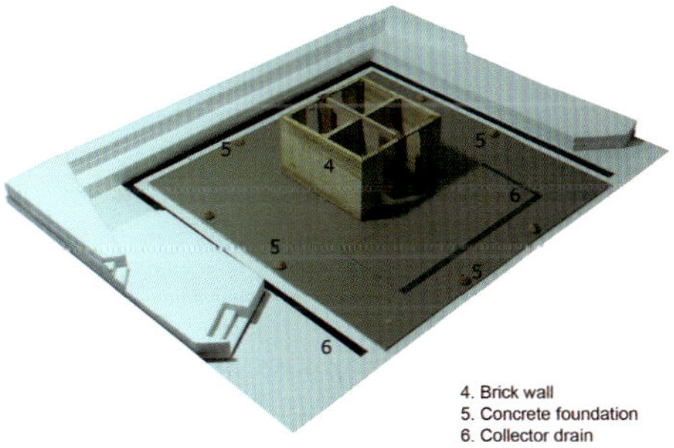

4. Brick wall
5. Concrete foundation
6. Collector drain

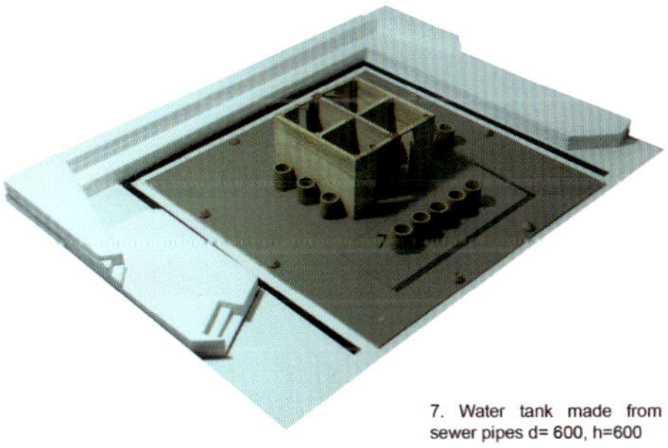

7. Water tank made from sewer pipes d= 600, h=600

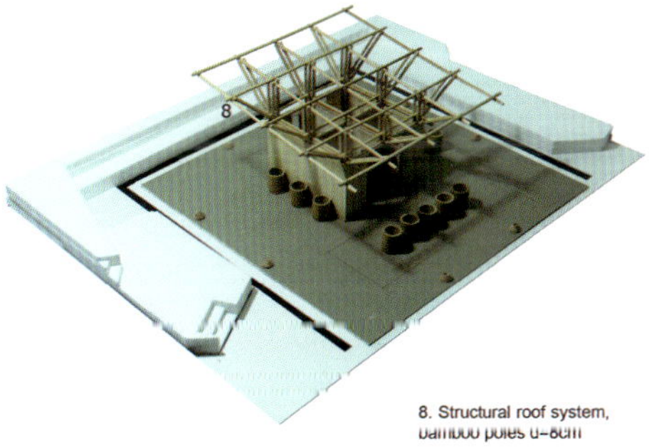

8. Structural roof system, bamboo poles d=8cm

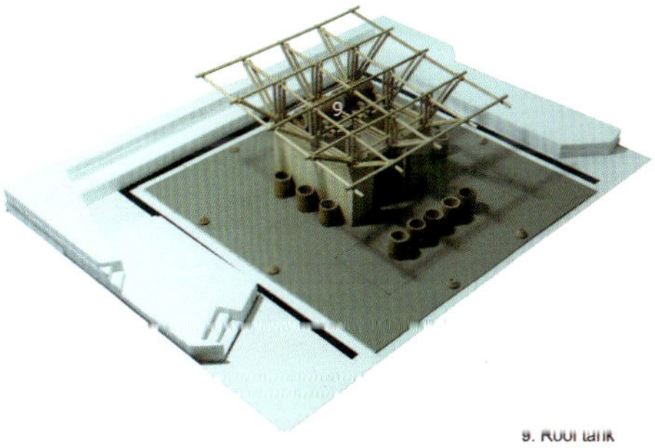

9. Roof tank

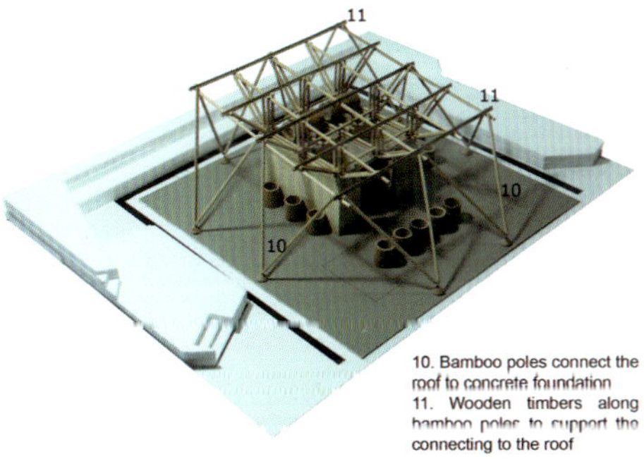

10. Bamboo poles connect the roof to concrete foundation
11. Wooden timbers along bamboo poles to support the connecting to the roof

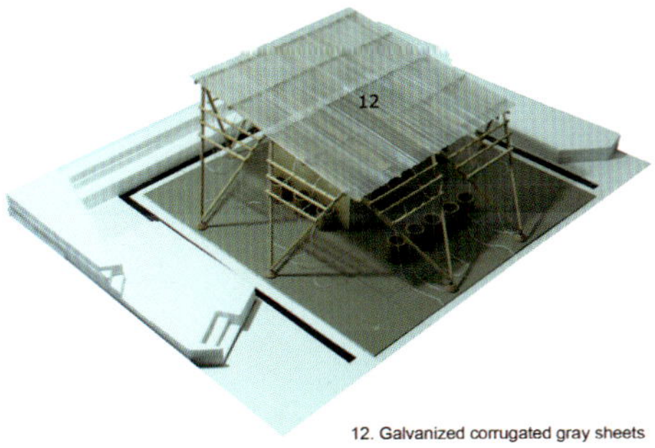

12. Galvanized corrugated gray sheets

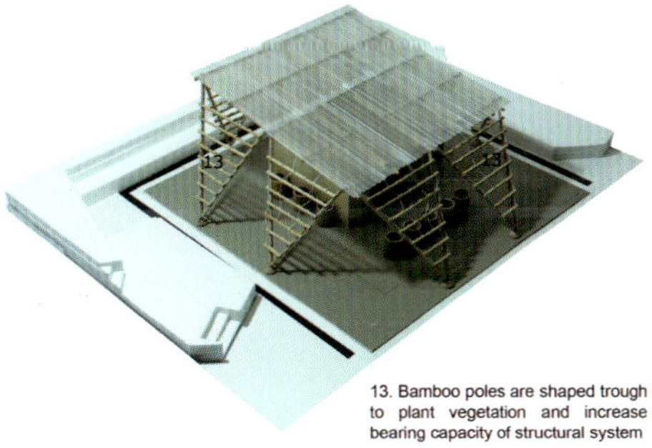

13. Bamboo poles are shaped trough to plant vegetation and increase bearing capacity of structural system

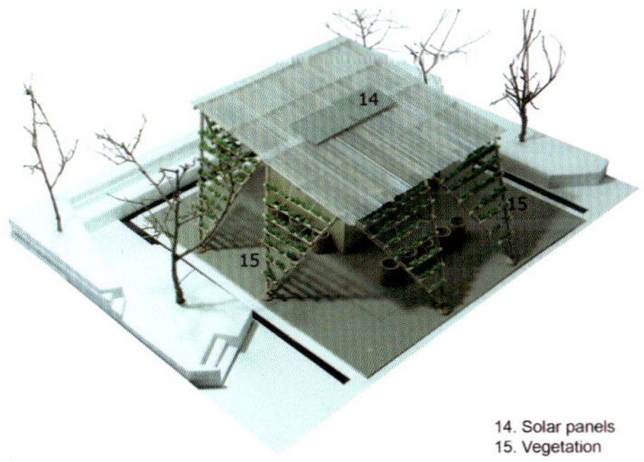

14. Solar panels
15. Vegetation

Construction Diagram

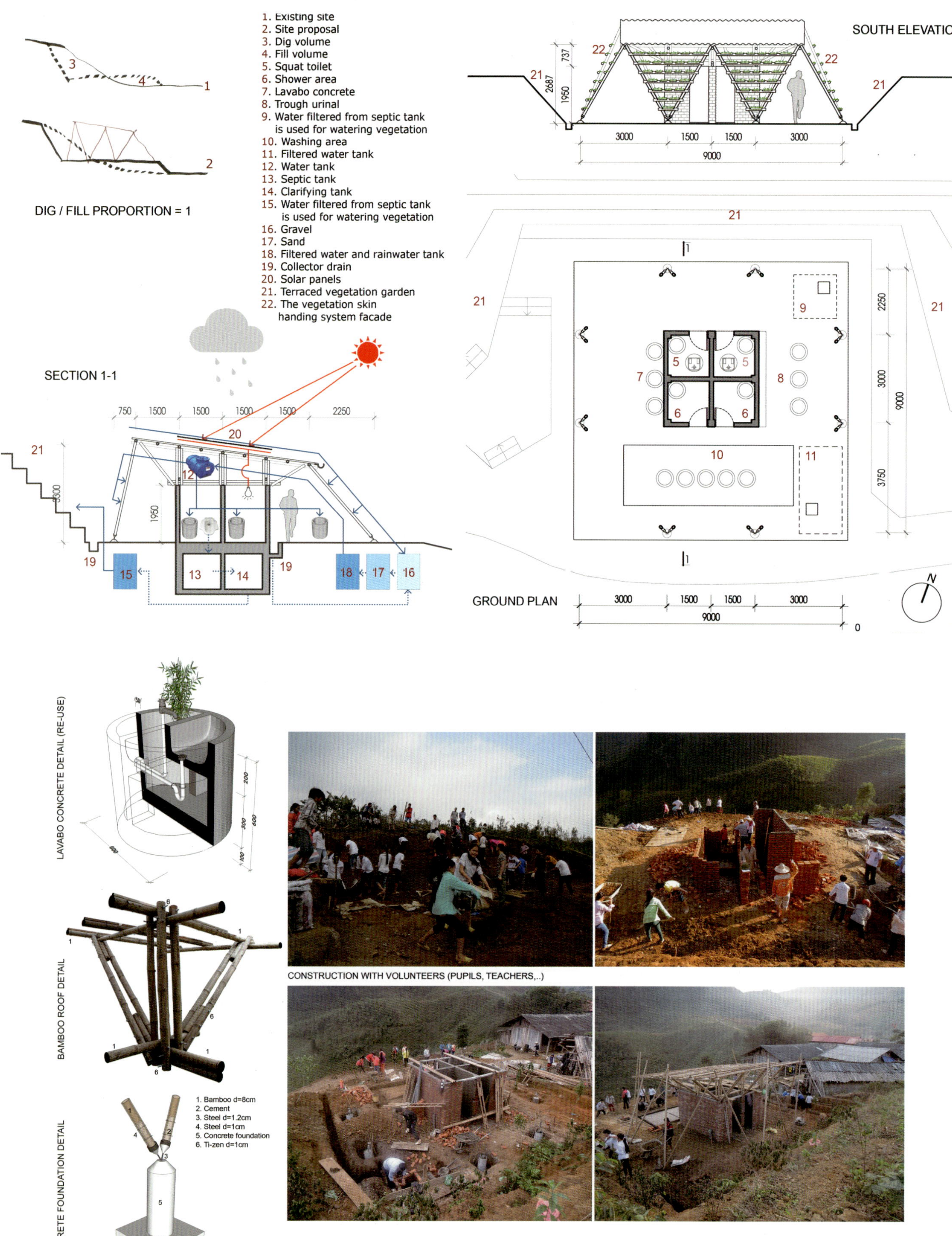

CONSTRUCTION WITH VOLUNTEERS (PUPILS, TEACHERS,..)

© Doan Thanh Ha

© Doan Thanh Ha

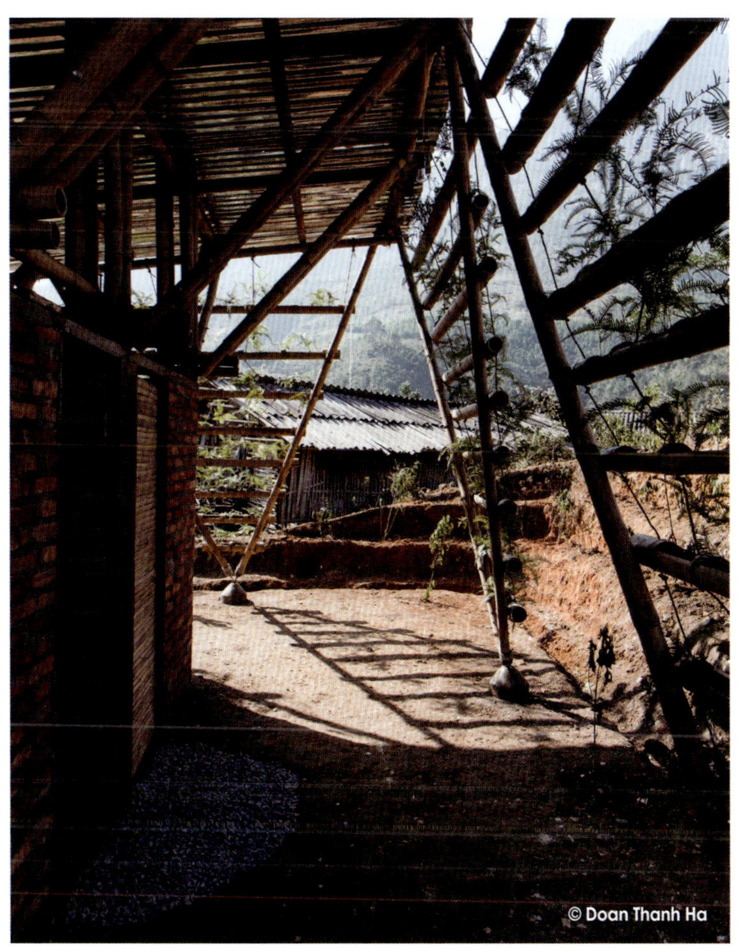

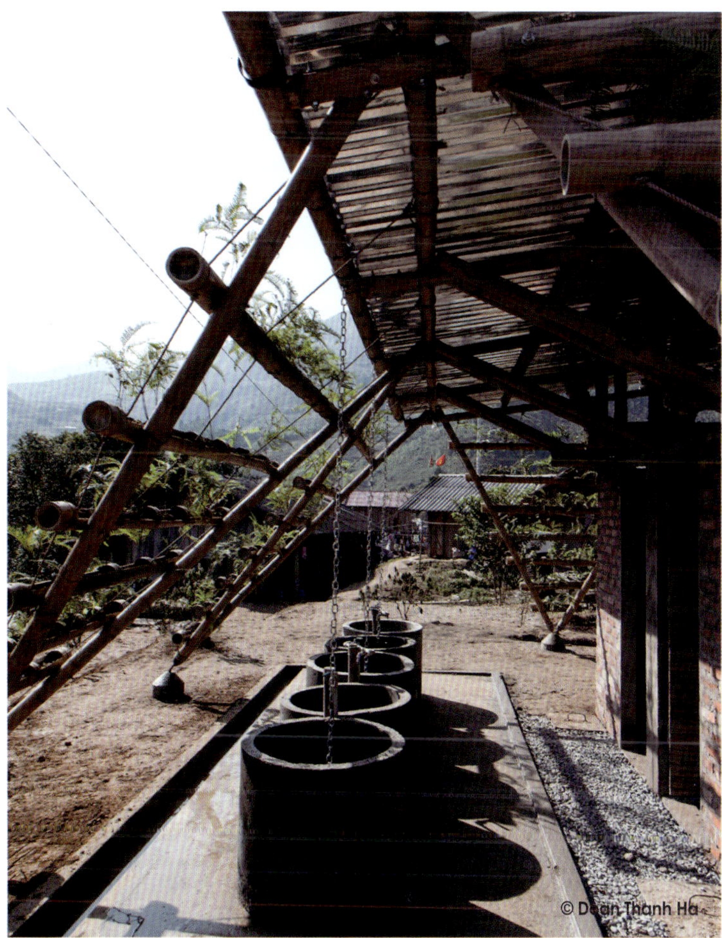

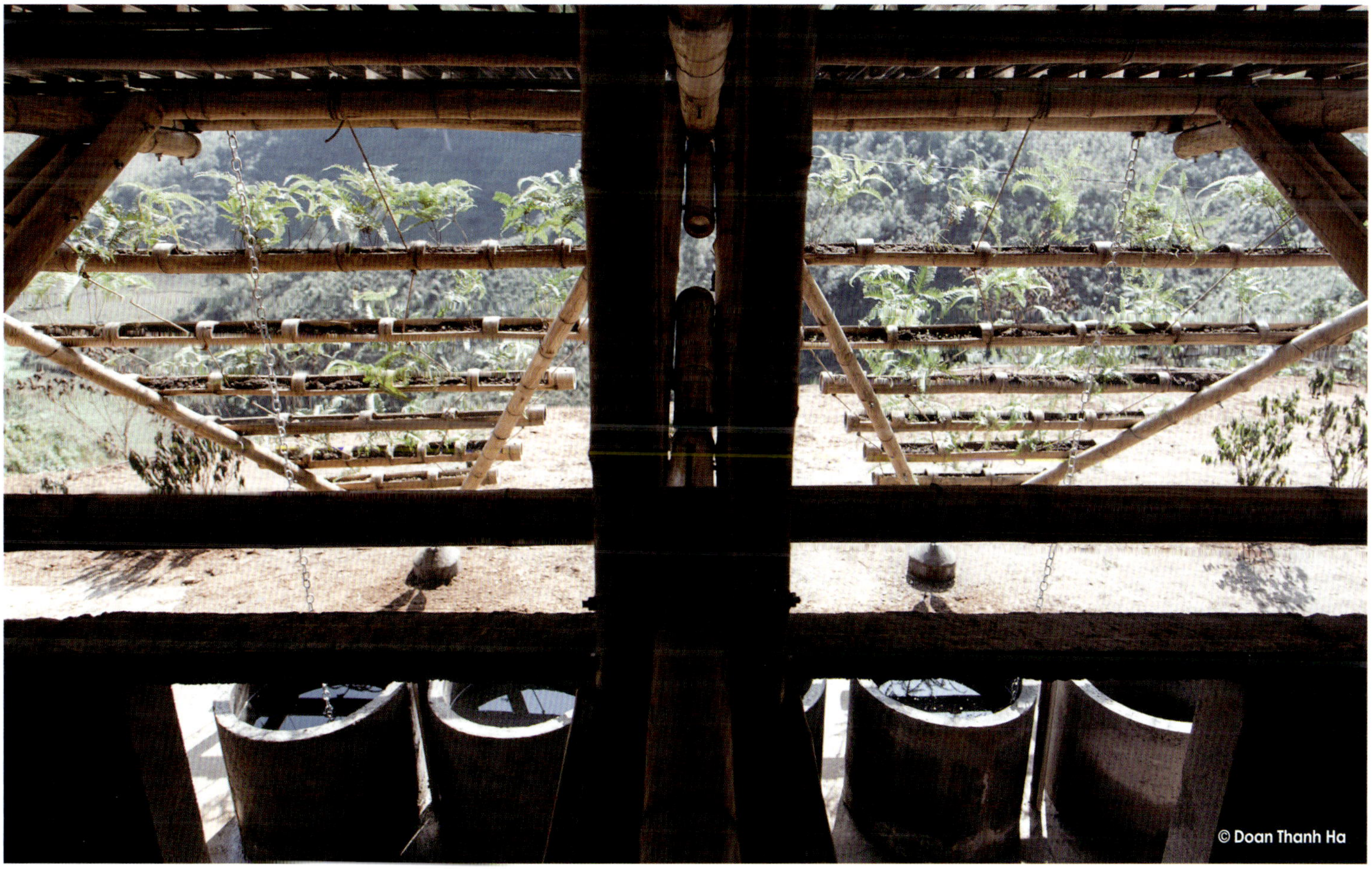
© Doan Thanh Ha

© Doan Thanh Ha

© Doan Thanh Ha

© Doan Thanh Ha

CANOPY: MoMA, New York, USA

nARCHITECTS

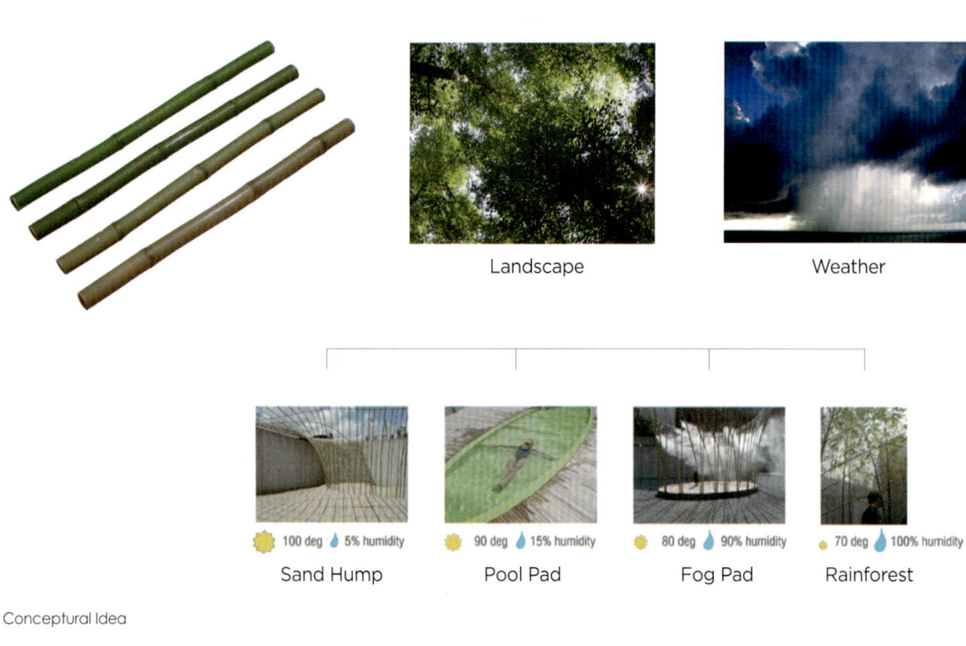

Conceptual Idea

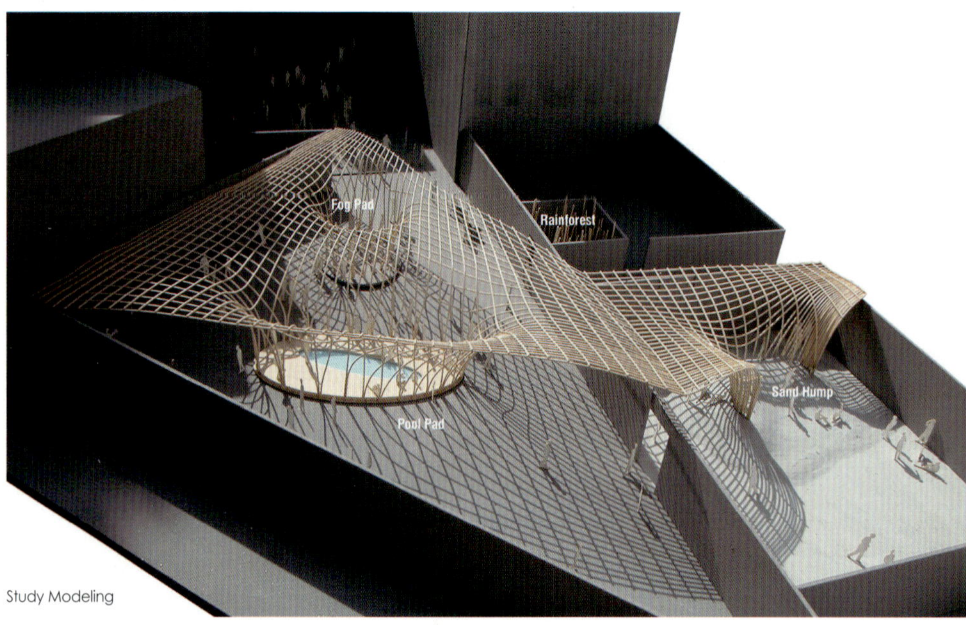

Study Modeling

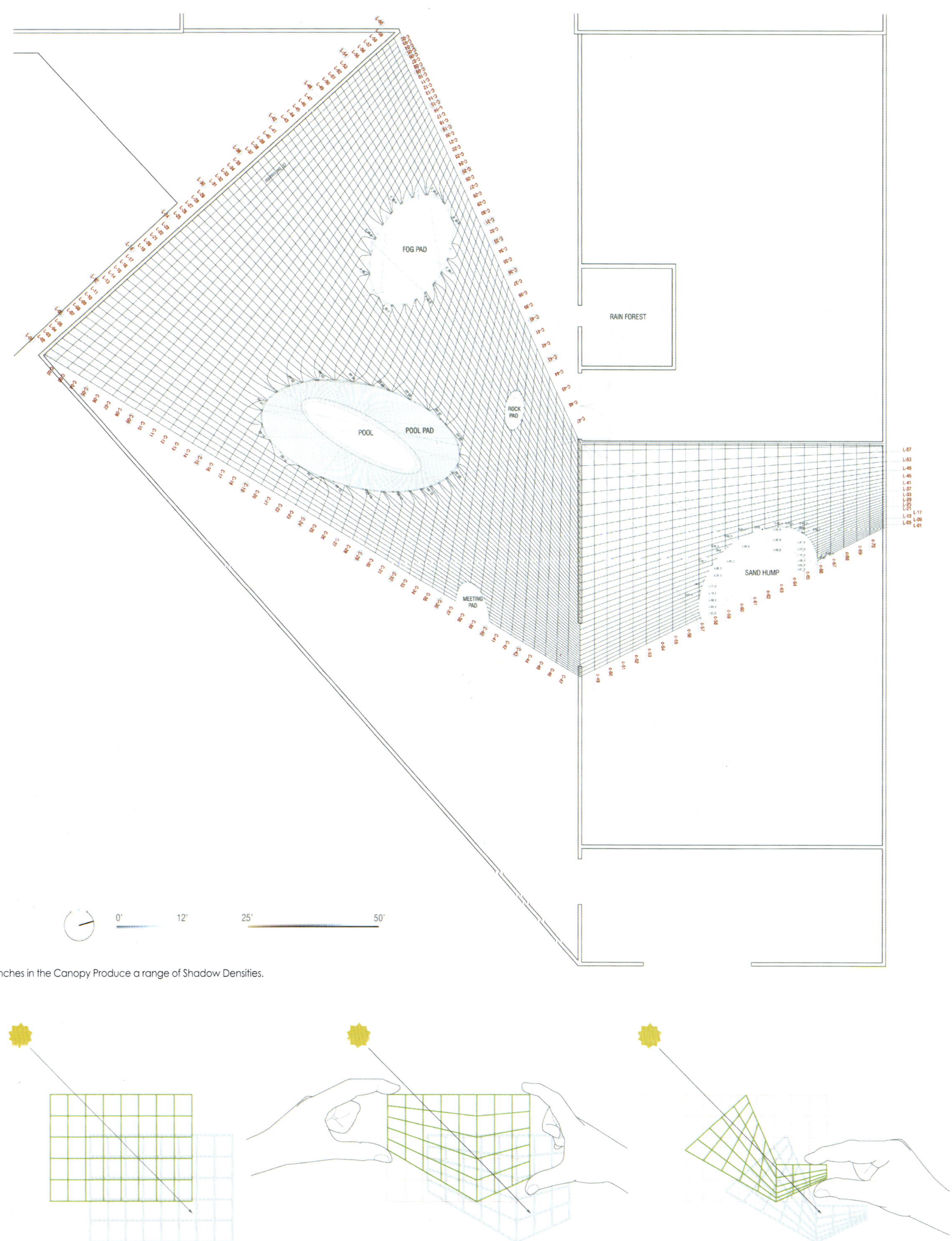

Pinches in the Canopy Produce a range of Shadow Densities.

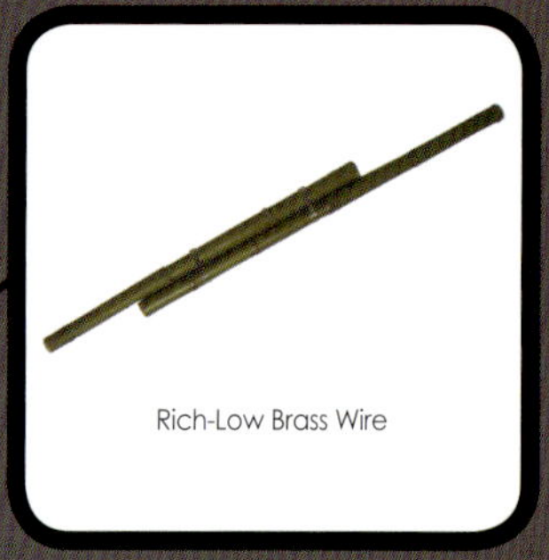

Rich-Low Brass Wire
Typical Overlap Splice

SPLICE TYPES

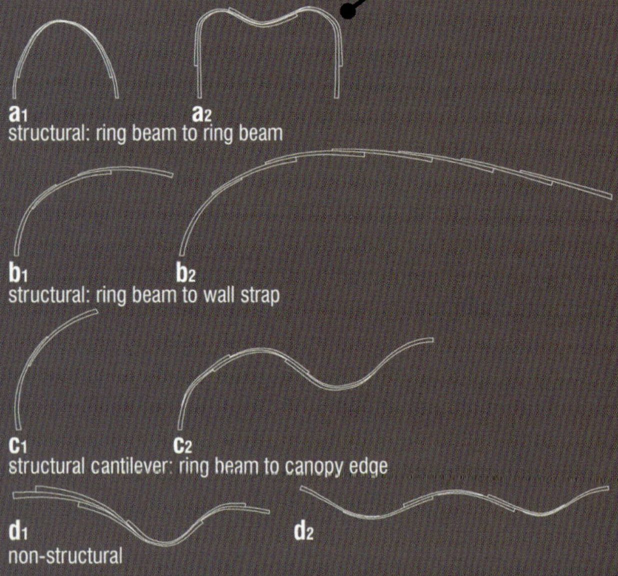

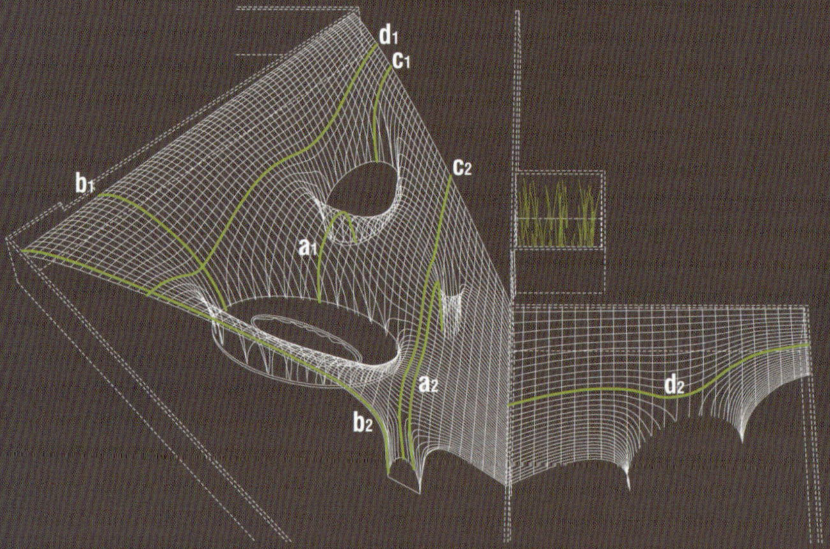

a1
a2
structural: ring beam to ring beam
b1
b2
structural: ring beam to wall strap
c1
c2
structural cantilever: ring beam to canopy edge
d1
non-structural
d2

d1
c1
b1
c2
a1
a2
b2
d2

Canopy juxtaposes the degree to which a natural material can be geometrically manipulated and controlled..

ERECTION SEQUENCE

Phase 1
Phase 2
Phase 3

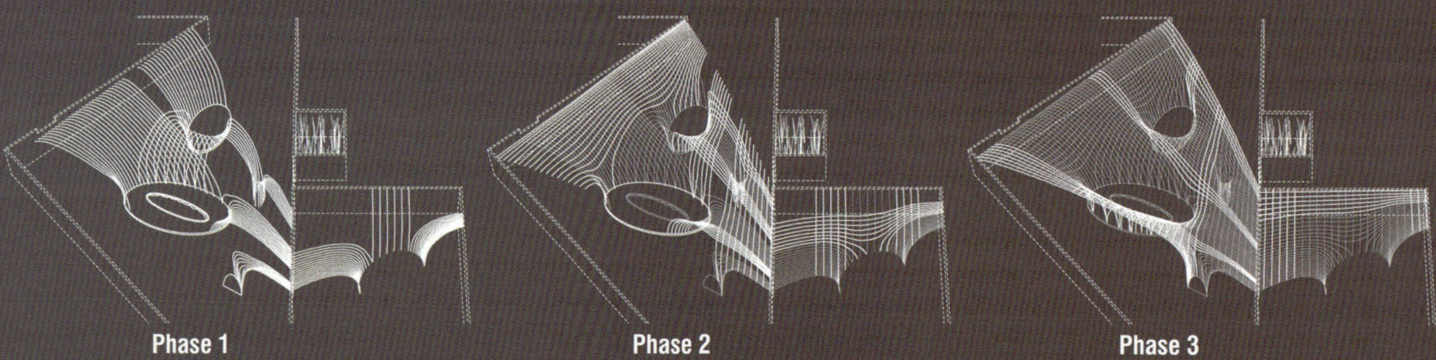

Construction Detai

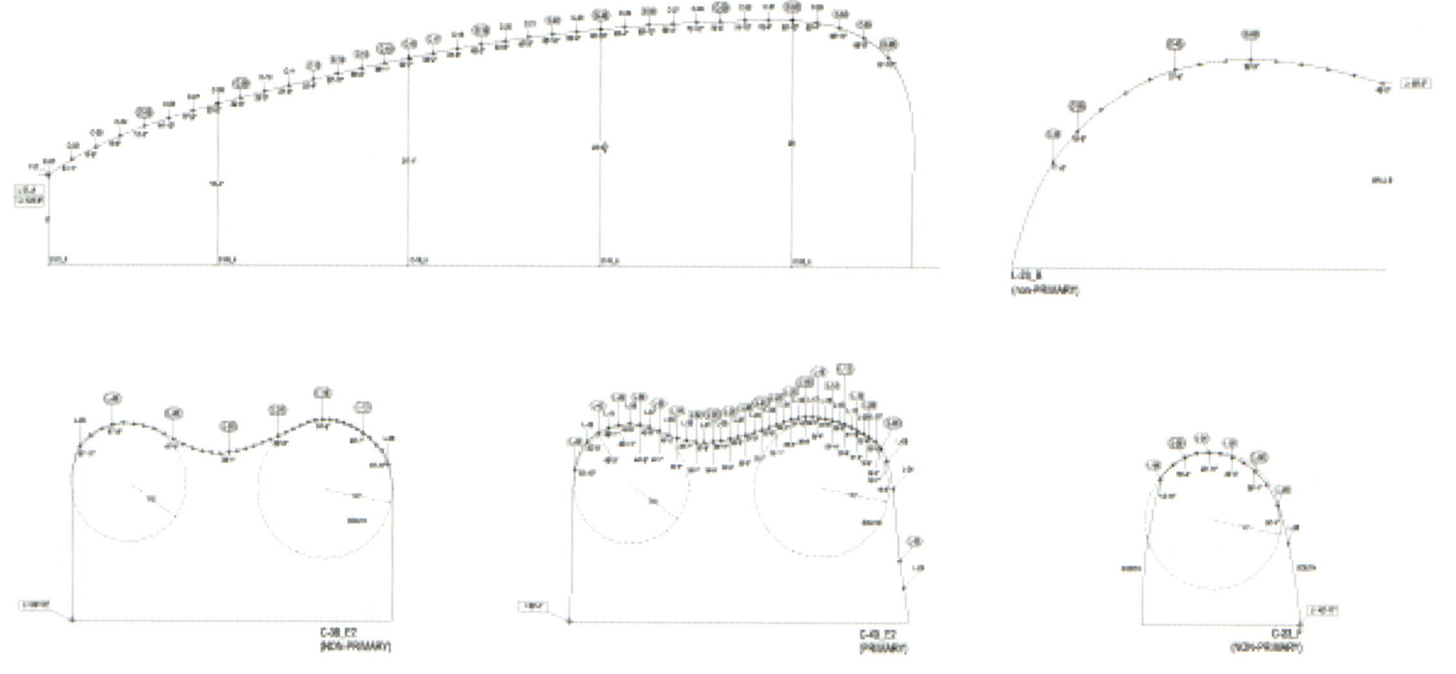

arc profiles

Arc Profiles

(SUB-PRIMARY)
(SUB-PRIMARY)

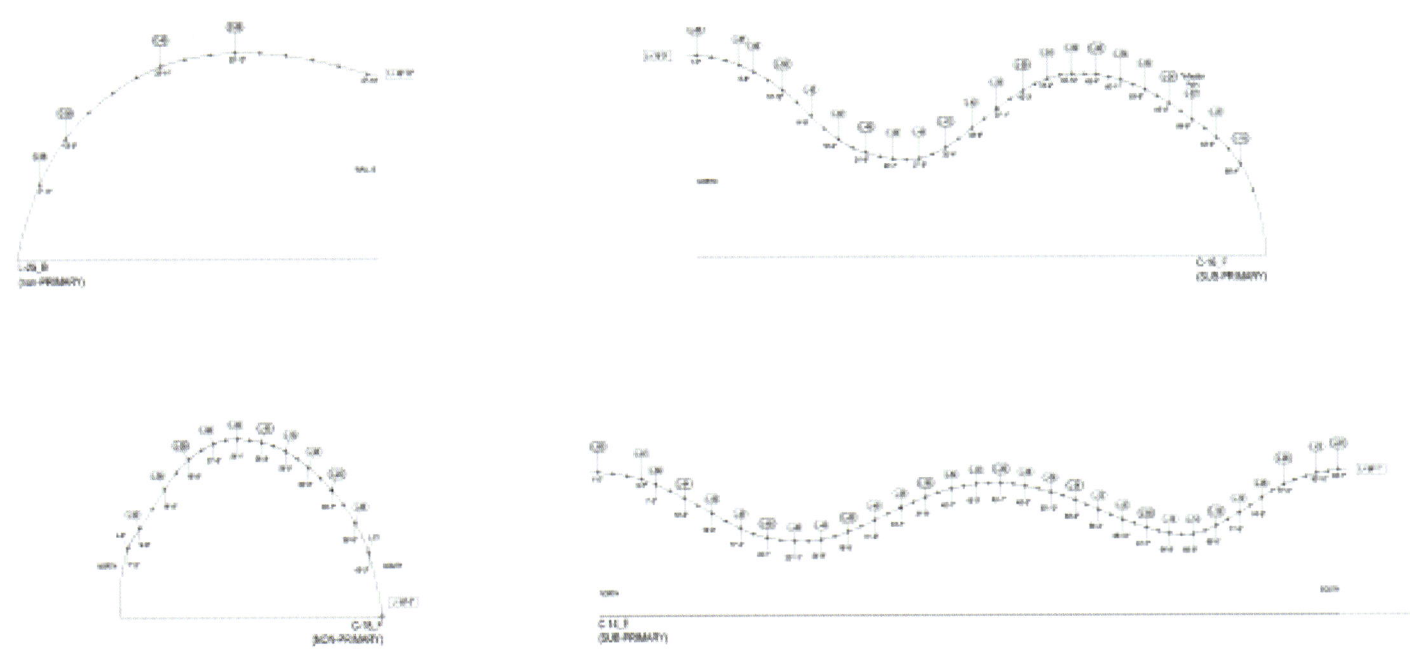

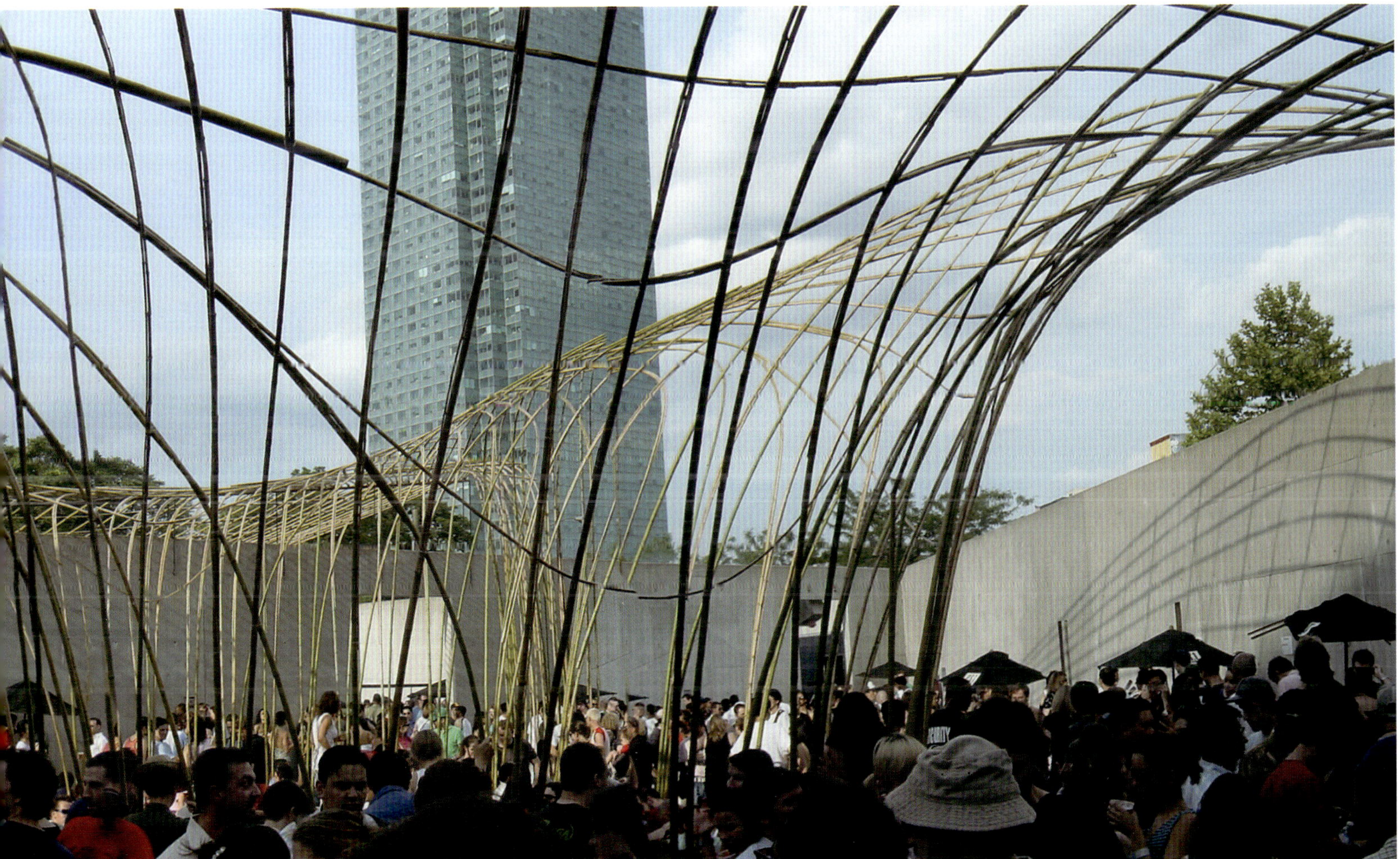

Construction Process

FKŻ QUARTER, Cracow, Poland
BudCud

© Haim Yafim Barbalat

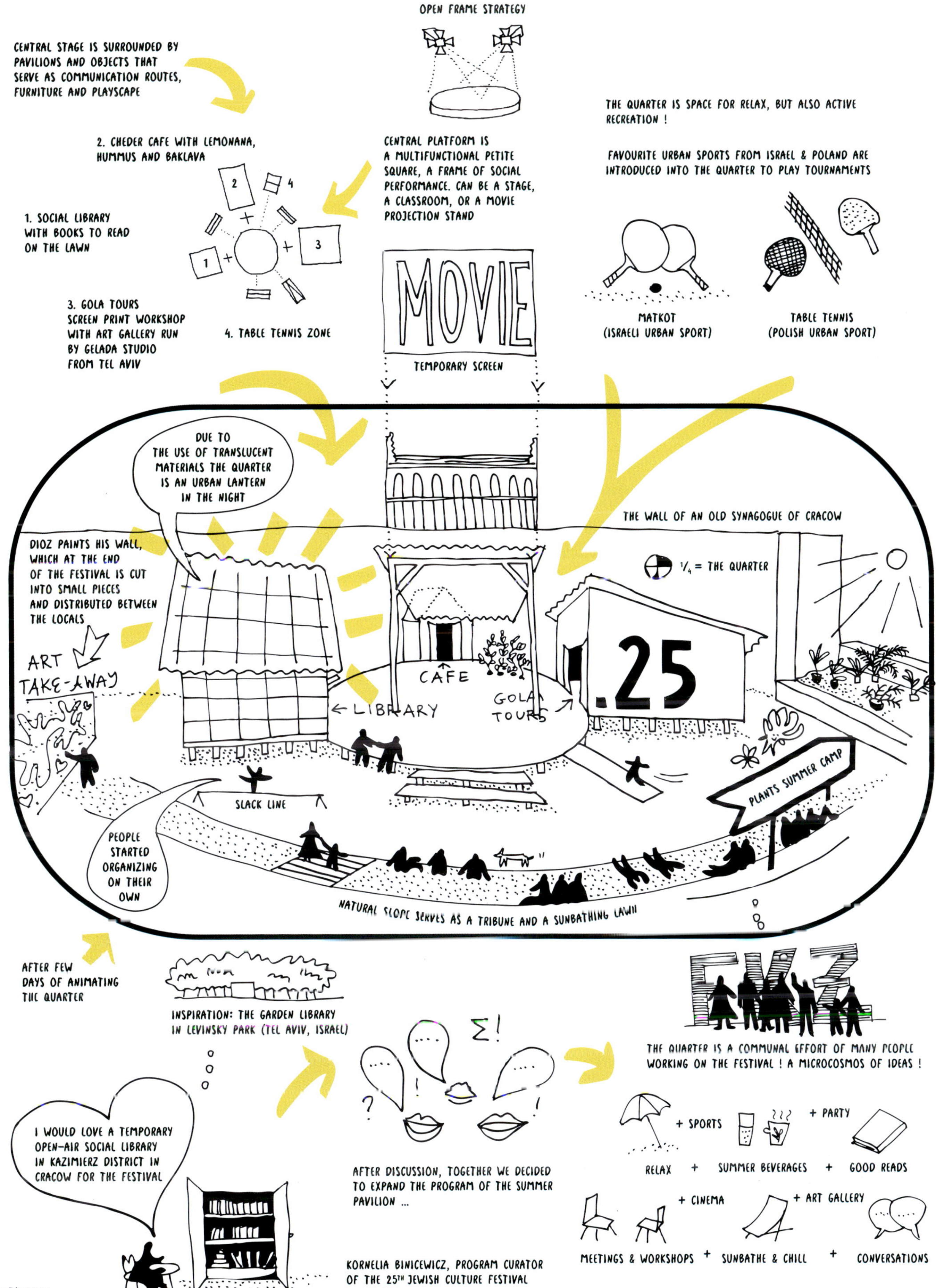

Program Diagram

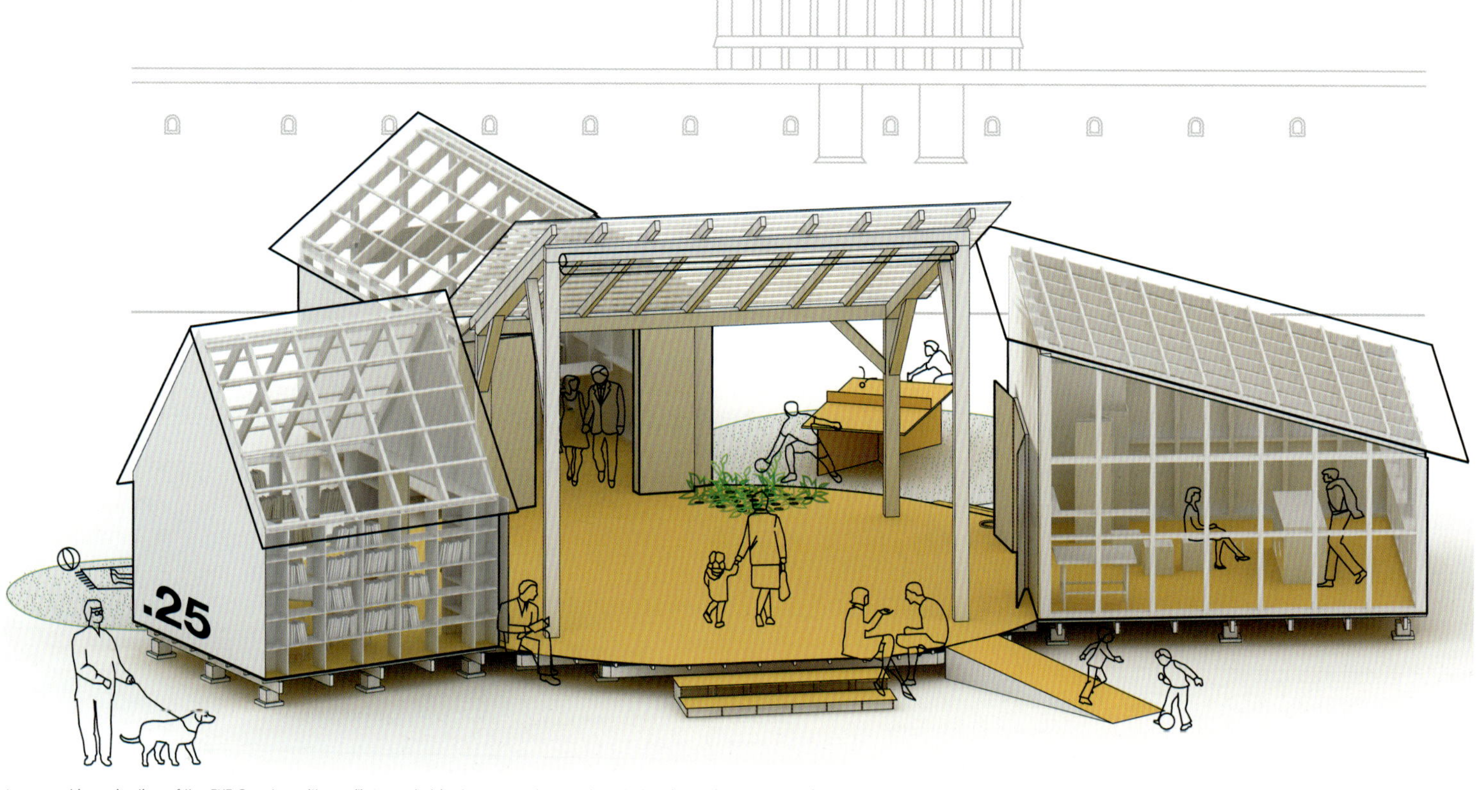

Axonometric projection of the FKZ Quarter, with pavilions and objects arranged around central social performance platform

© Haim Yafim Barbalat

© Haim Yafim Barbalat

© Haim Yafim Barbalat

© Haim Yafim Barbalat

© Haim Yafim Barbalat

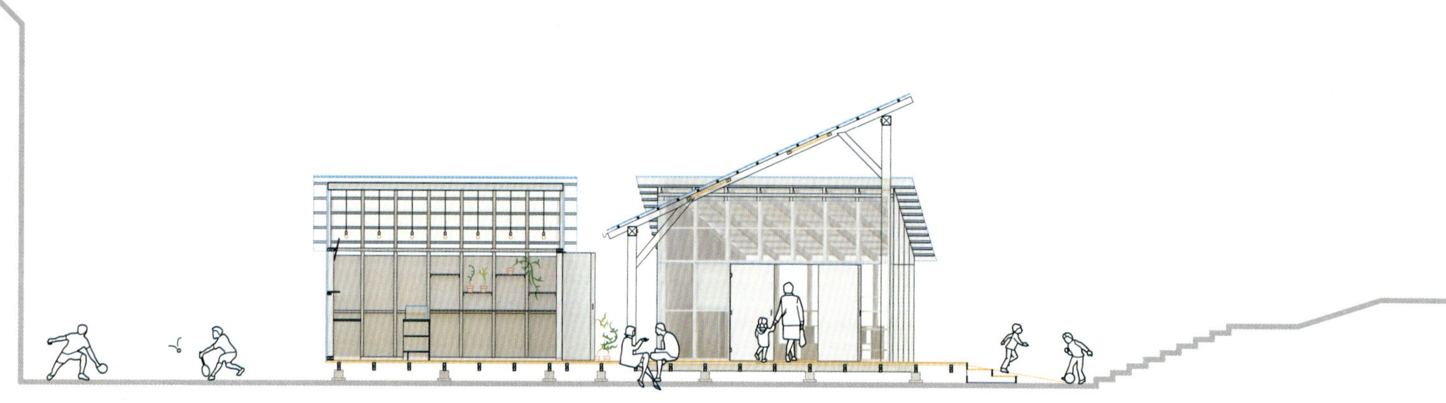

Section

© Haim Yafim Barbalat

© Haim Yafim Barbalat

NON STOP
PARKING
STRZEŻONY
24h

© Haim Yafim Barbalat

Kaap Skil, Maritime and Beachcombers Museum, Texel, the Netherland

mecanoo

Facade

TEXEL
NB 5
OL

04
06

KEELUNG, Keelung, Taiwan

Guallart Architects

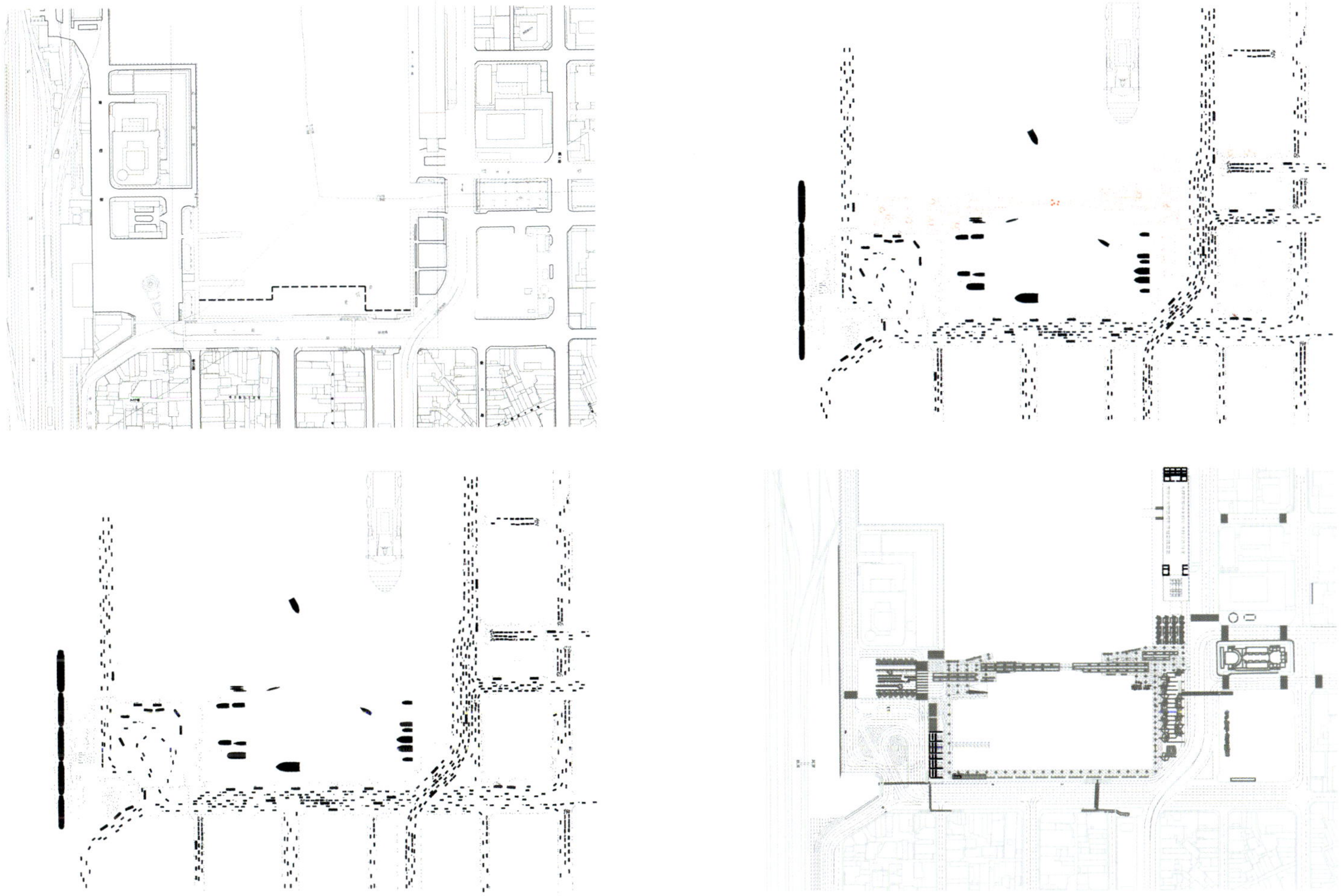

Analysis

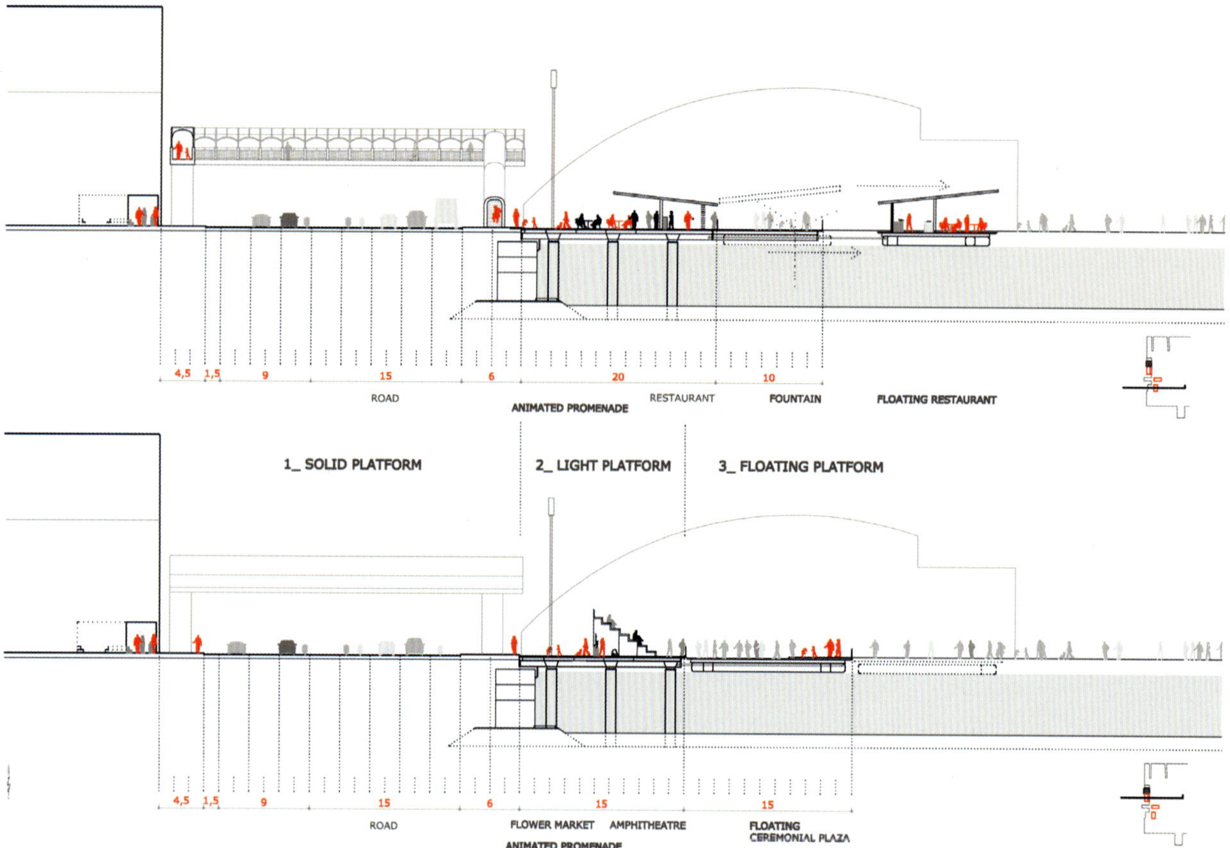

Platform Section

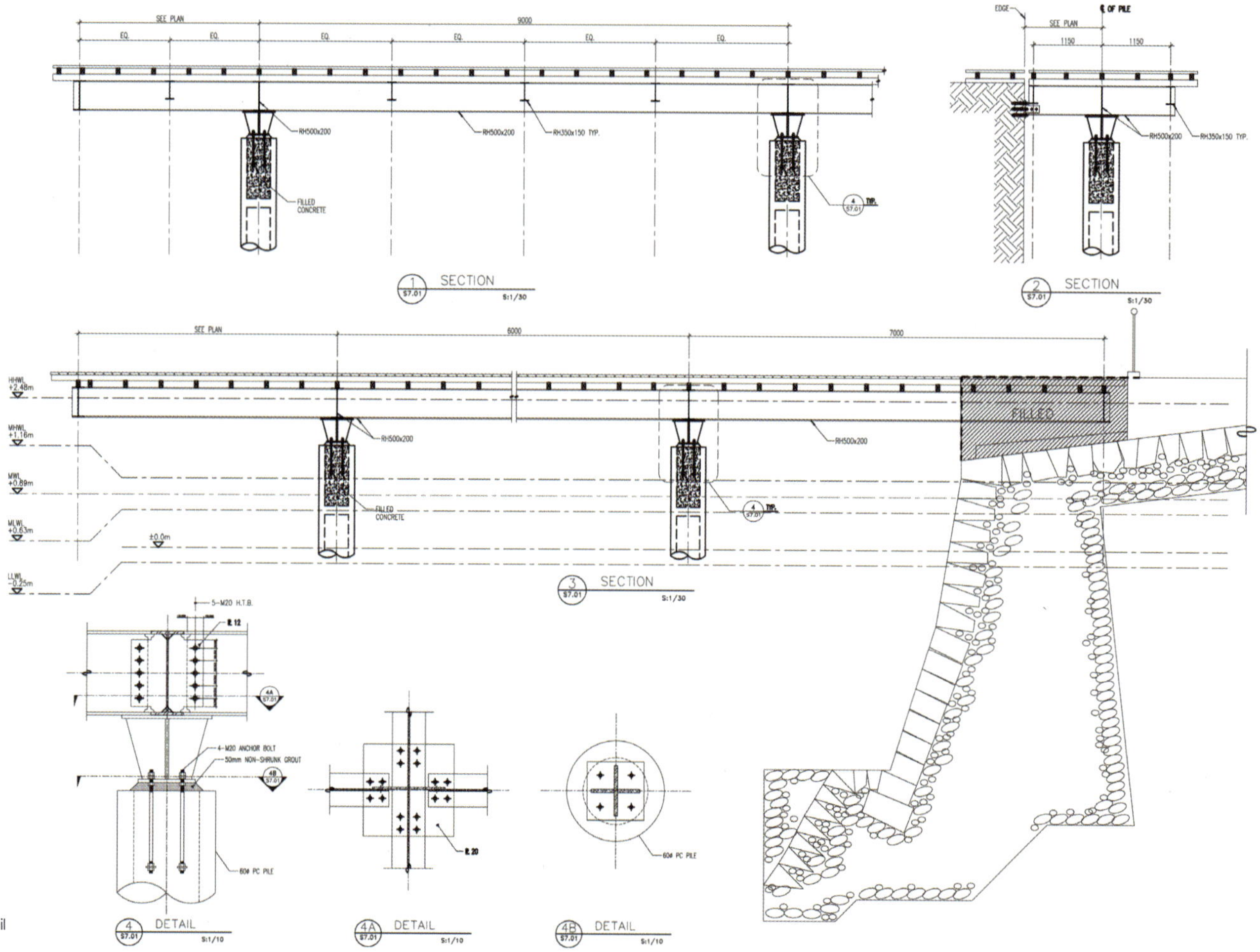

SECTION
SECTION
SECTION
FILLED
DETAIL
DETAIL
DETAIL

Platform Detail

Study Modeling

MUR-Mobile Unit Reference, Post Disaster Sites
b4architects

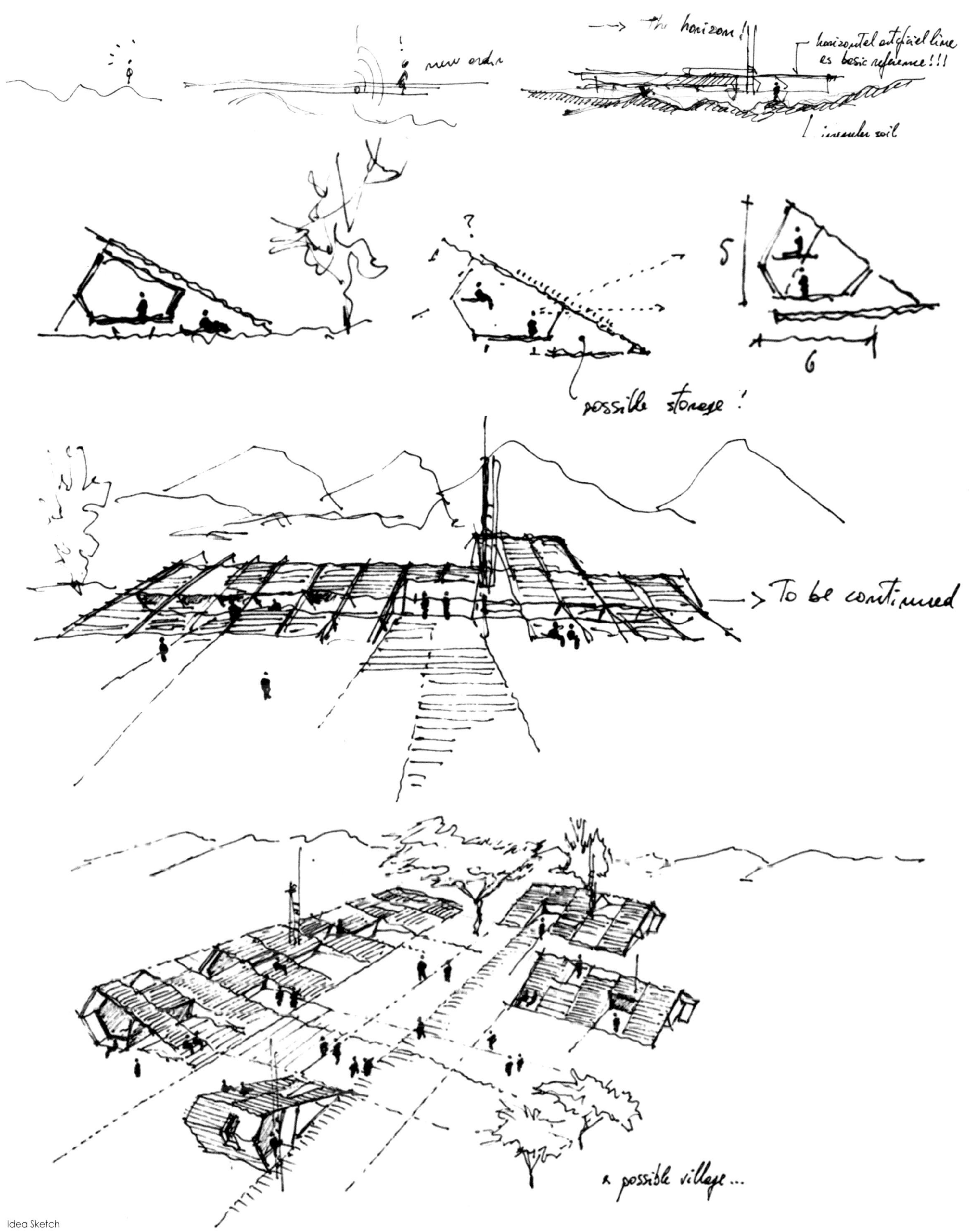
the horizon!
new order
5
6
possible storage!
To be continued
a possible village...

Idea Sketch

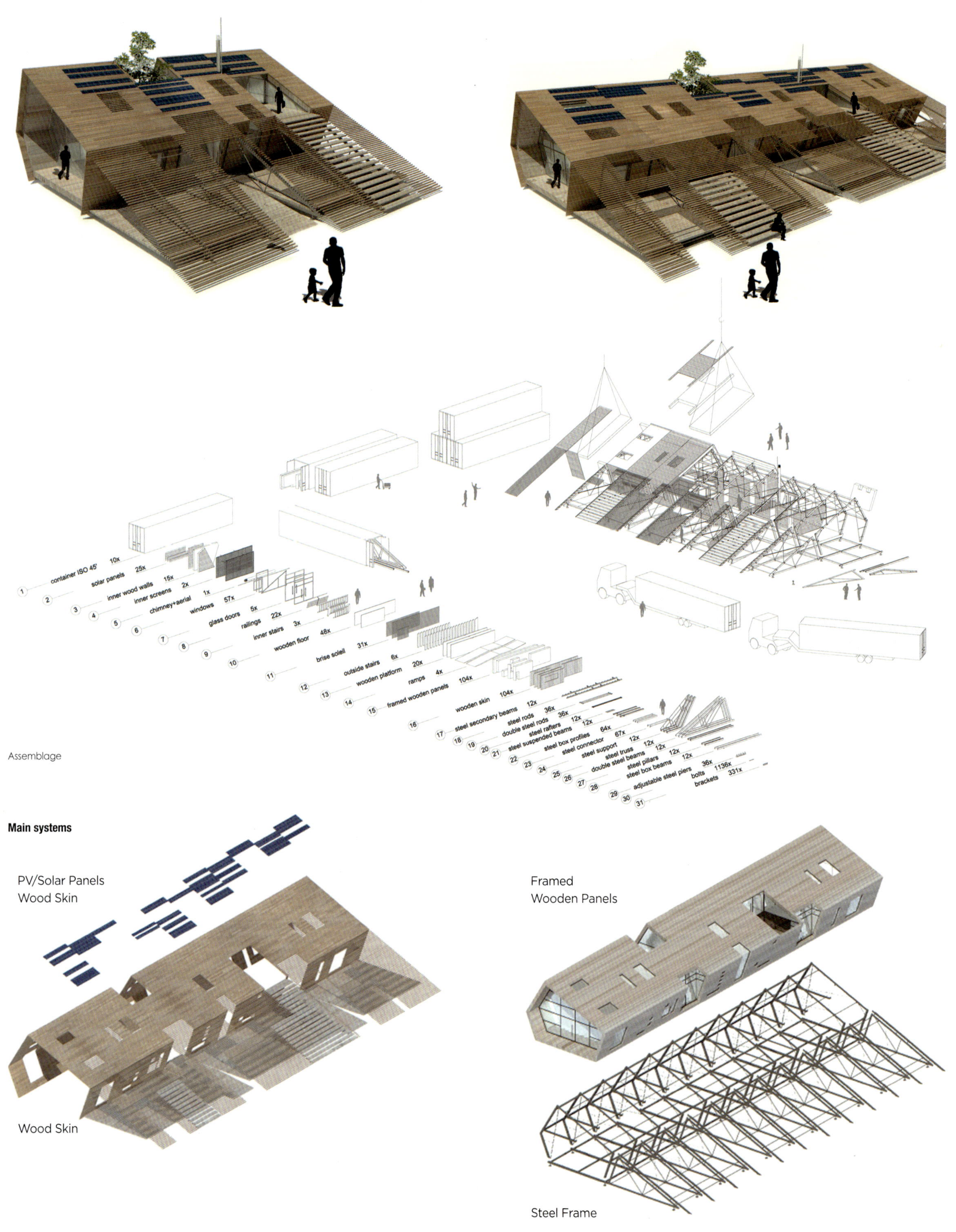

Assemblage

Main systems

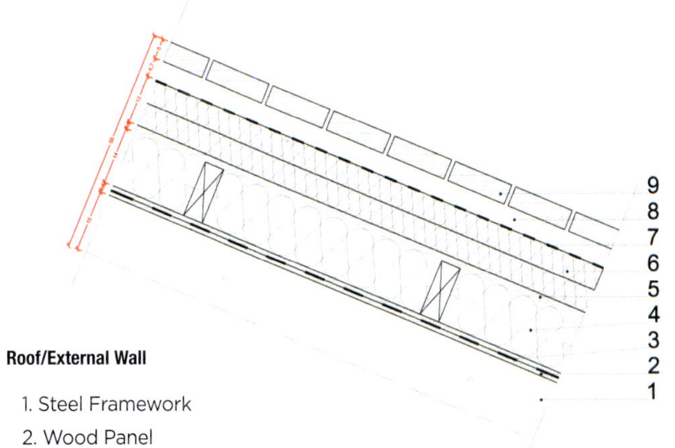

Roof/External Wall

1. Steel Framework
2. Wood Panel
3. Membrances
4. Insulation: Fiber Wood 50kg/mc
5. Wood Panel
6. Insulation: Fiber Wood 160kg/mc
7. Waterproofing Membrances
8. Air Gap
9. External Wood

A_L	U_i	A_i	F_i	$R_{si}+R_{se}$	A_i*U_i*F_i
m^2	W/m^2K	m^2			W/K
286.00	0.14	258.94	1.00	0.20	36.25
286.00		**258.94**			**36.25**

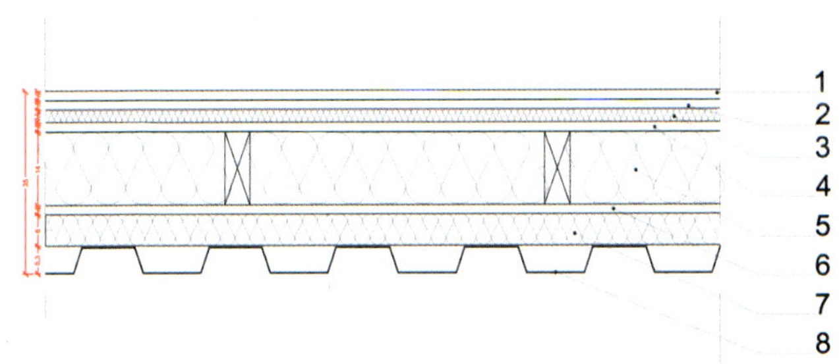

Floor

1. Wooden Floor
2. Plasterboard
3. Glass Fiber
4. Wood Panel
5. Insulation: Fiber Wood 50kg/mc
6. Wood Panel
7. Insulation: Fiber Wood 160kg/mc
8. Corrugated Sheet

A_L	U_i	A_i	F_i	$R_{si}+R_{se}$	A_i*U_i*F_i
m^2	W/m^2K	m^2			W/K
219.00	0.15	219.00	1.00	0.34	32.85
219.00		**219.00**			**32.85**

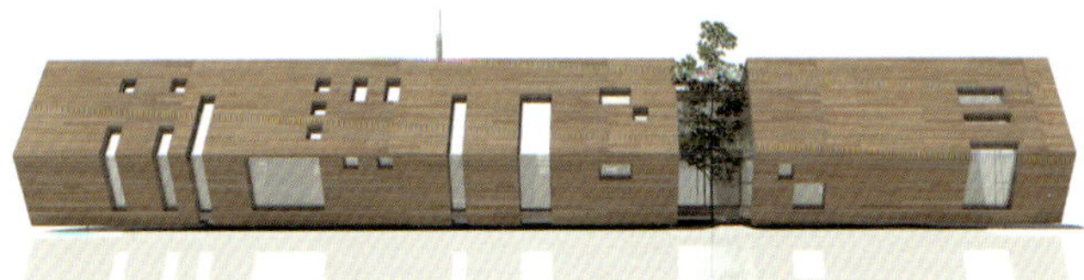

New Bus Station in Drammen, Drammen, Norway

studio grön arkitekter ab

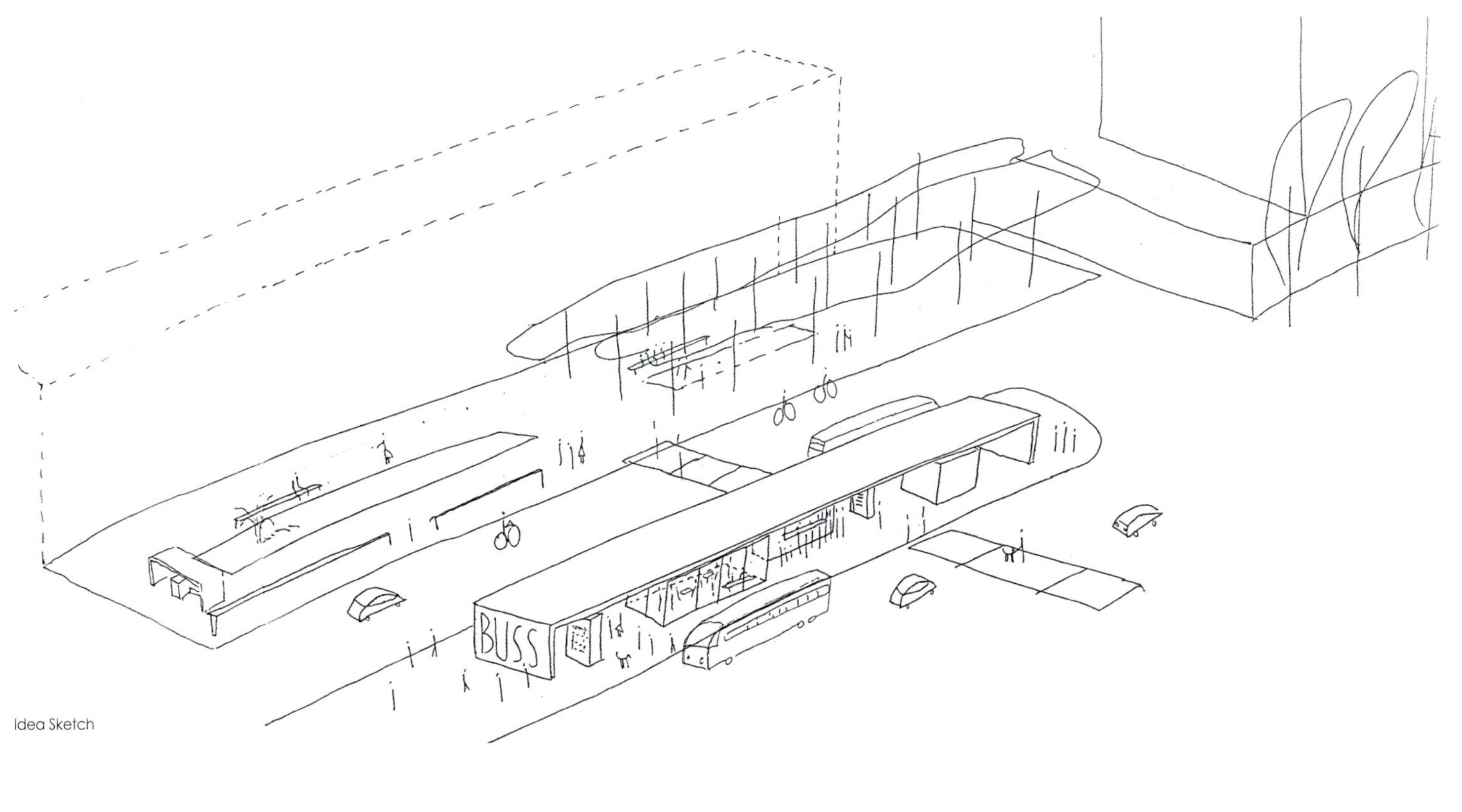

Idea Sketch

Idea Sketch Elevation

© Devegg Ruud

© Devegg Ruud

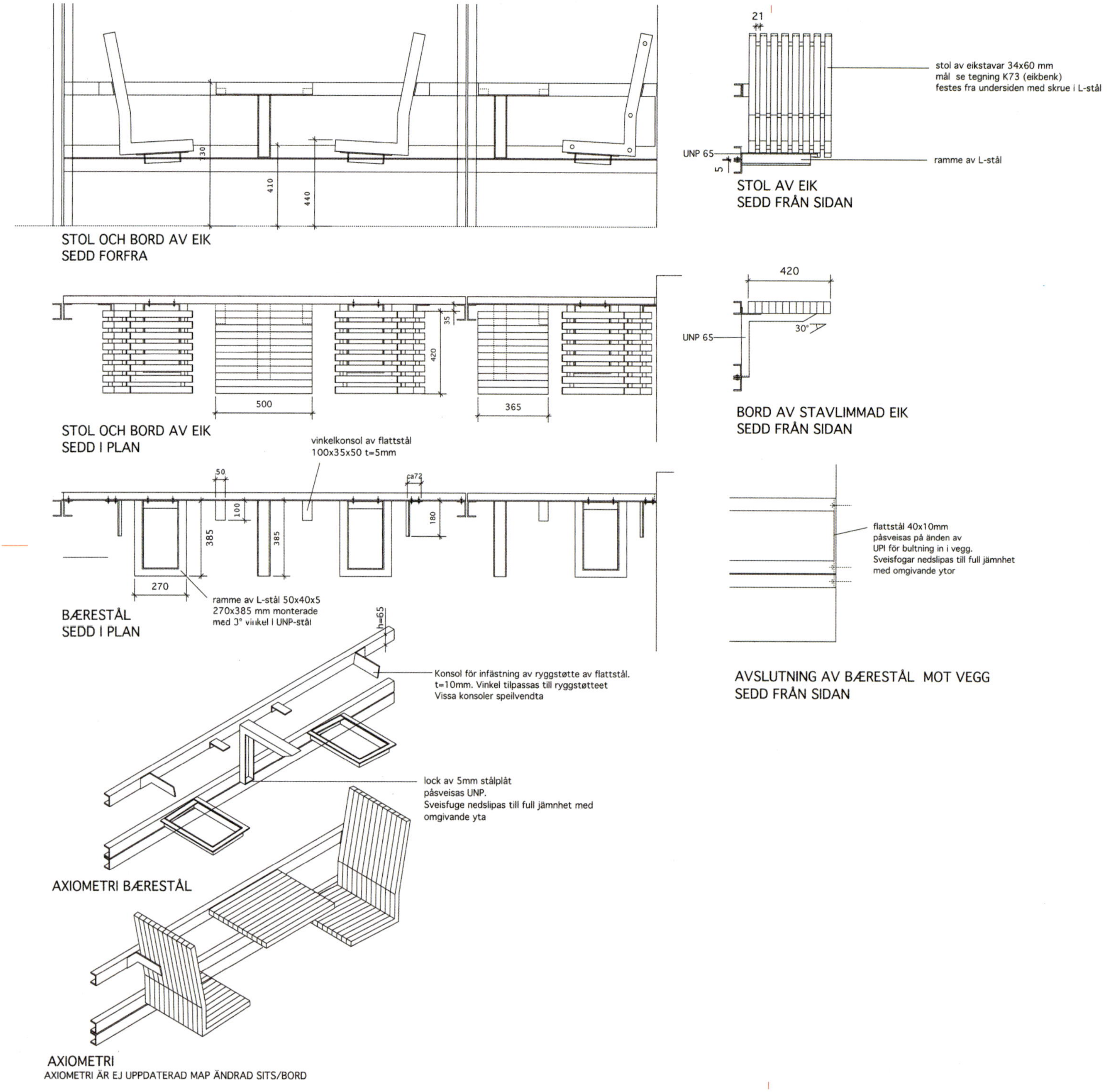

STOL OCH BORD AV EIK
SEDD FORFRA

21

stol av eikstavar 34x60 mm
mål se tegning K73 (eikbenk)
festes fra undersiden med skrue i L-stål

UNP 65

ramme av L-stål

STOL AV EIK
SEDD FRÅN SIDAN

500

365

420

35

STOL OCH BORD AV EIK
SEDD I PLAN

420

30°

UNP 65

BORD AV STAVLIMMAD EIK
SEDD FRÅN SIDAN

vinkelkonsol av flattstål
100x35x50 t=5mm

50

ca72

100

180

385

385

270

ramme av L-stål 50x40x5
270x385 mm monterade
med 3° vinkel i UNP-stål

BÆRESTÅL
SEDD I PLAN

flattstål 40x10mm
påsveisas på änden av
UPI för bultning in i vegg.
Sveisfogar nedslipas till full jämnhet
med omgivande ytor

AVSLUTNING AV BÆRESTÅL MOT VEGG
SEDD FRÅN SIDAN

h=65

Konsol för infästning av ryggstøtte av flattstål.
t=10mm. Vinkel tilpassas till ryggstøtteet
Vissa konsoler speilvendta

lock av 5mm stålplåt
påsveisas UNP.
Sveisfuge nedslipas till full jämnhet med
omgivande yta

AXIOMETRI BÆRESTÅL

AXIOMETRI
AXIOMETRI ÄR EJ UPPDATERAD MAP ÄNDRAD SITS/BORD

Detail

Facade

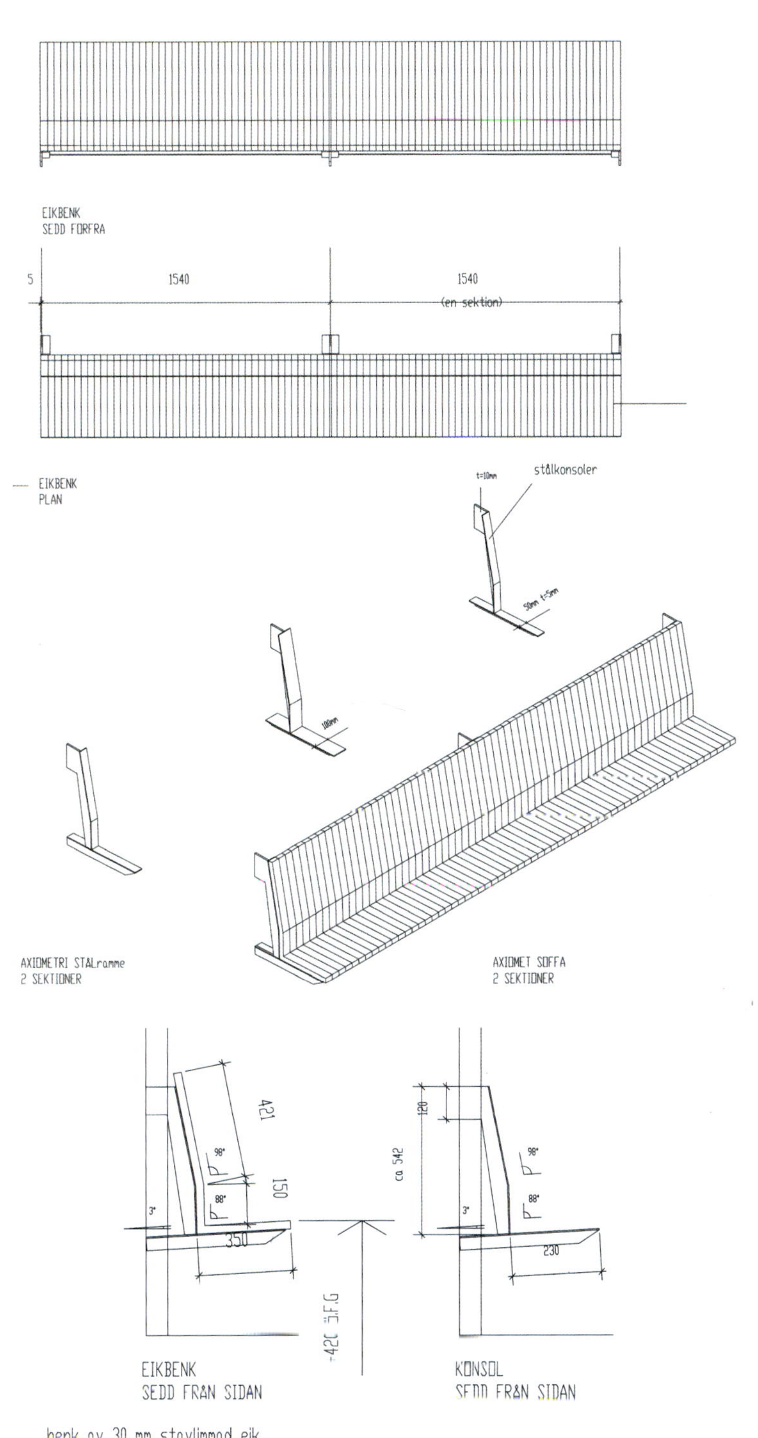
EIKBENK
SEDD FORFRA
5
1540
1540
(en sektion)
EIKBENK
PLAN
stålkonsoler
AXIOMETRI STÅLramme
2 SEKTIONER
AXIOMET SOFFA
2 SEKTIONER
421
150
350
ca 542
120
230
EIKBENK
SEDD FRÅN SIDAN
KONSOL
SEDD FRÅN SIDAN
benk av 30 mm stavlimmad eik
riktning enligt tegning.
Bredde eilstav 35mm

Drammen Kommune, Byprosjekter / Reinertsen Engineering as / busstasjon Strømsø
studio grön arkitekter ab januari 2003

P
C

© Devegg Ruud

PIAVE, Belluno,Italy

Archi[te]n-sions

Serralves Exhibition Pavilion

murmuro

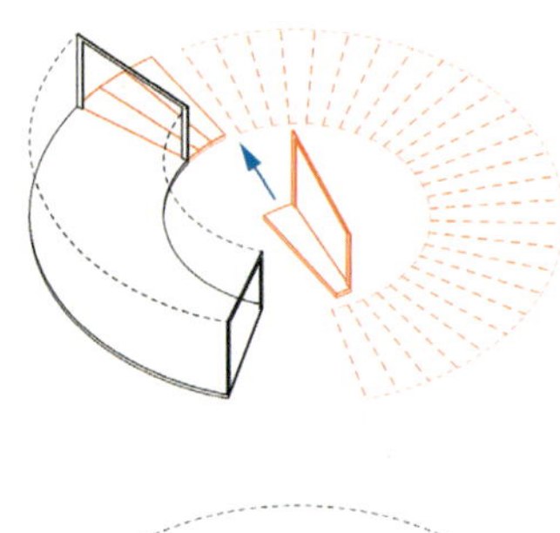
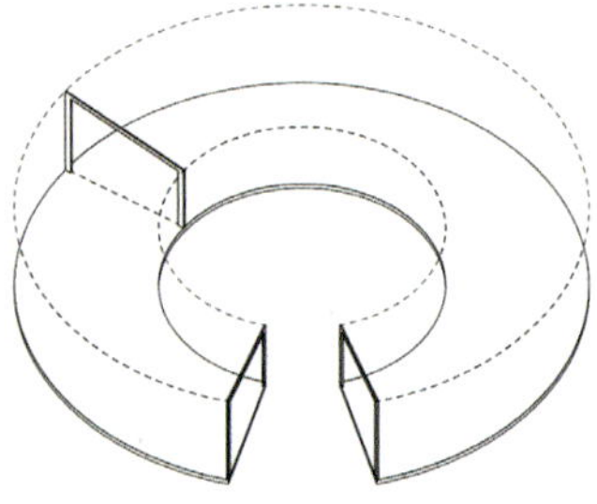

Construction - Module

Modular system in radial configuration. Ant module has a floor area of 12.5 sqm

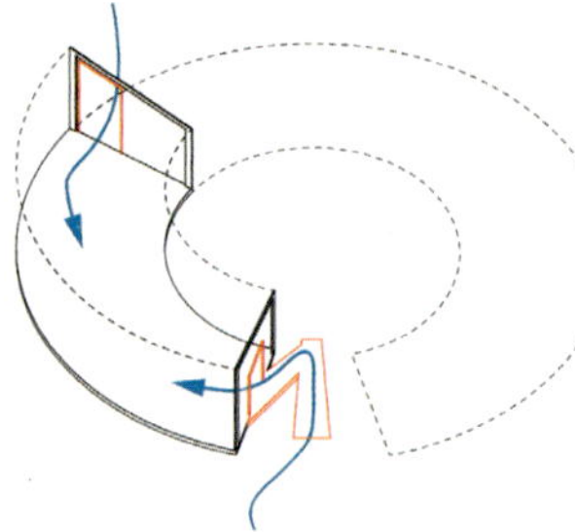
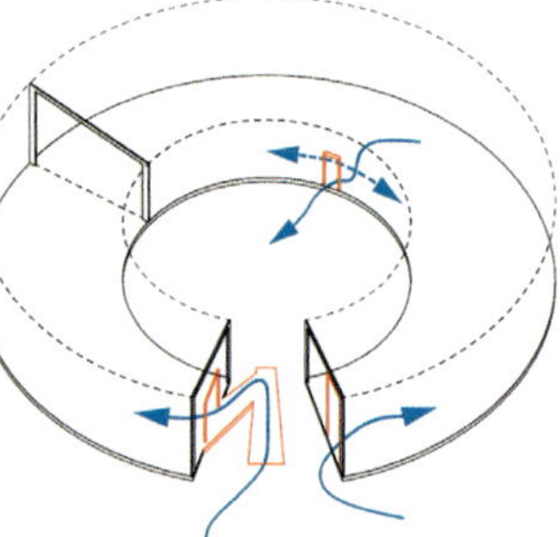

Access

Located at opposite ends for the public and service / logistics. Inner face may include optional location entries to access the outside or serve as an emergency exit

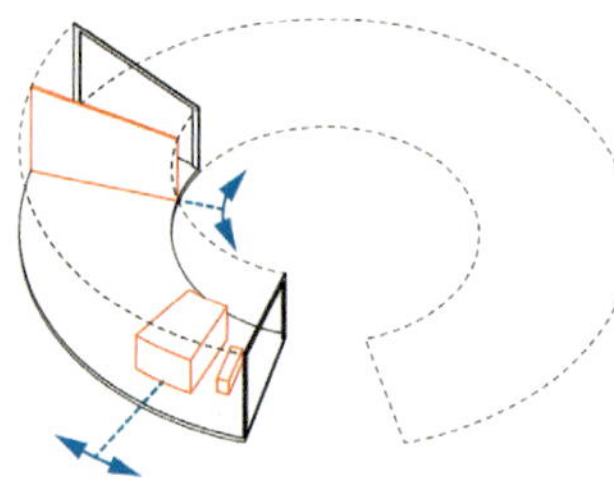
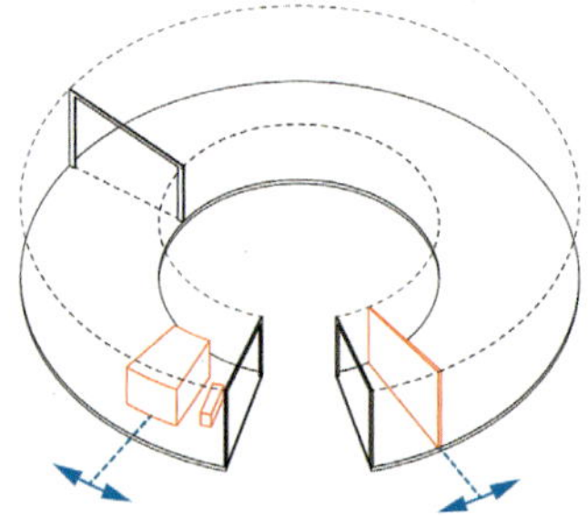

Exceptional modules

With variable location for reception areas (counter, cloakroom, wc) and service / logistics according to needs

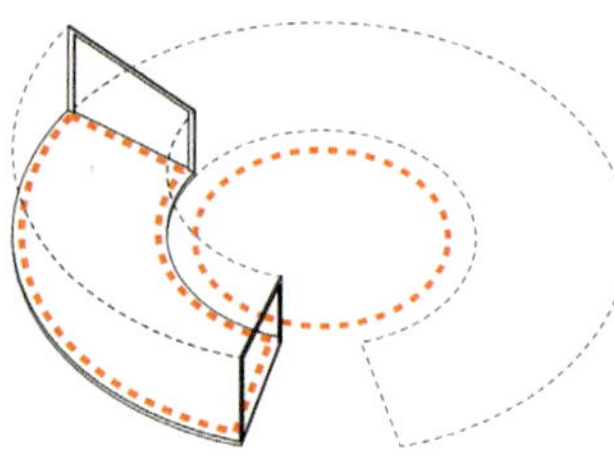
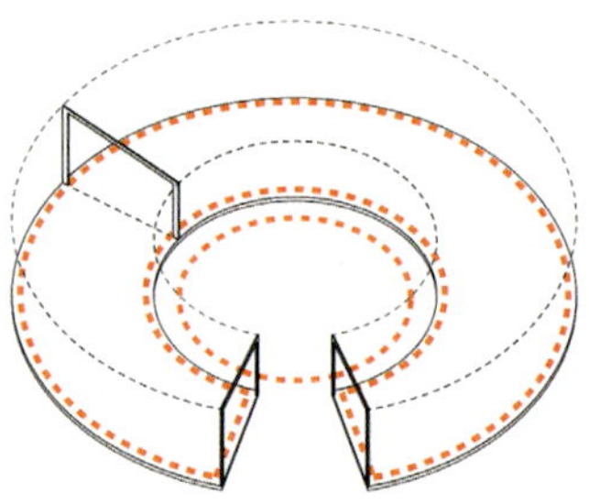

Exhibition space

Area configured between the exceptional modules that can grow towards the outside

Study Modeling

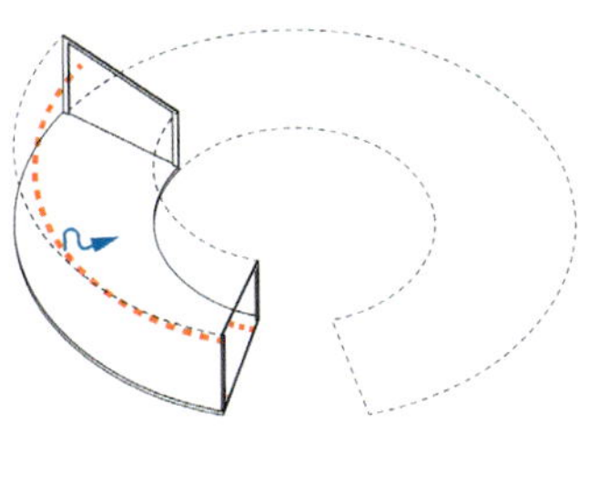
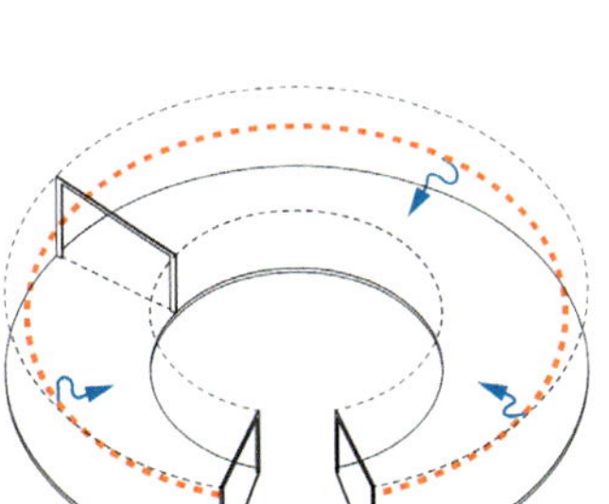
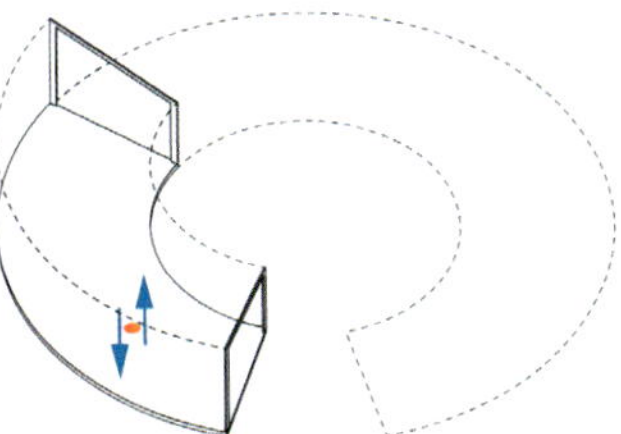
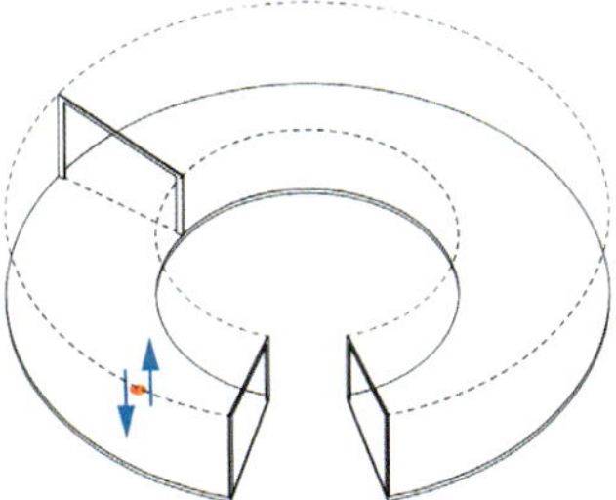
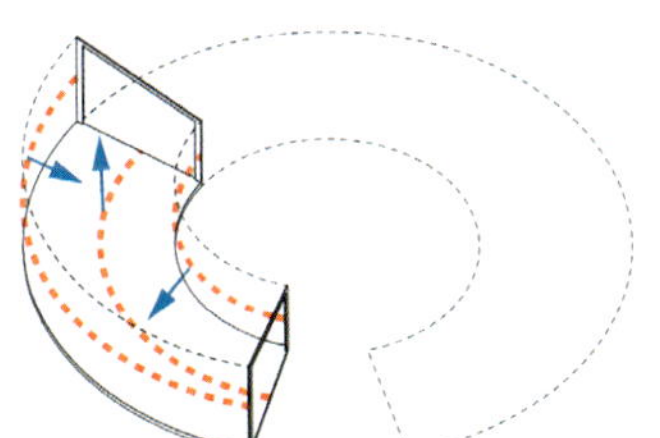
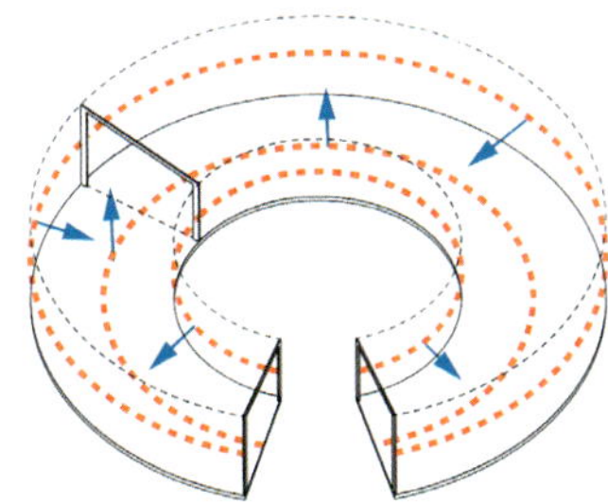

Air conditioning

Inclusion of a airduct in textile sleeve extending throught all the length of the pavilion

Plumbing

The supply and caption point for sanitary waters will serve the sanitary installation

Electrical feed lines

Present all along the pavilion both on floor and walls at different heights

Diagrams

200sqm - 600sqm Pavilions

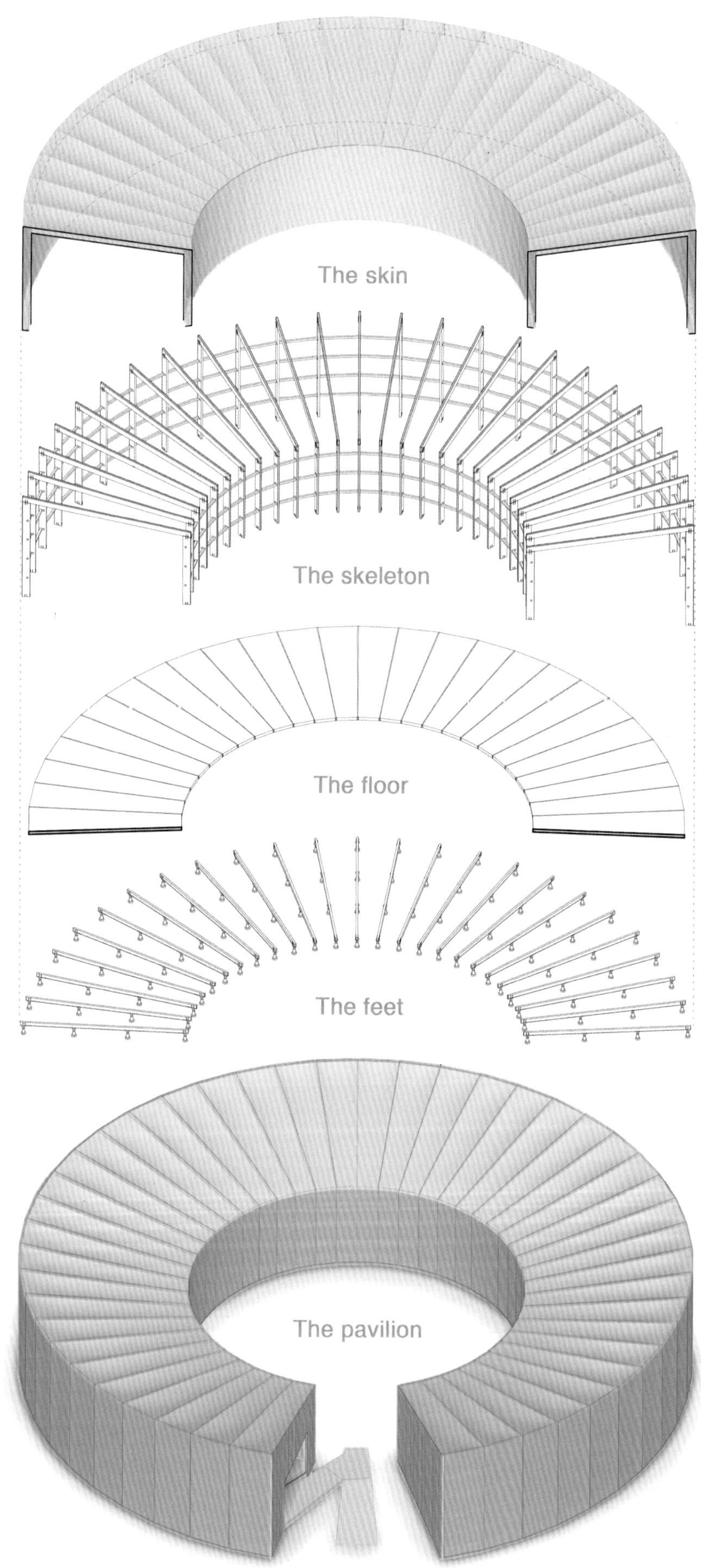

Components Axonometric

Exterior View

Interior View

Courtyard View

Vinaròs Project, Vinaròs, Spain
Guallart Architects

© Laura Cantarella, Nuria Díaz

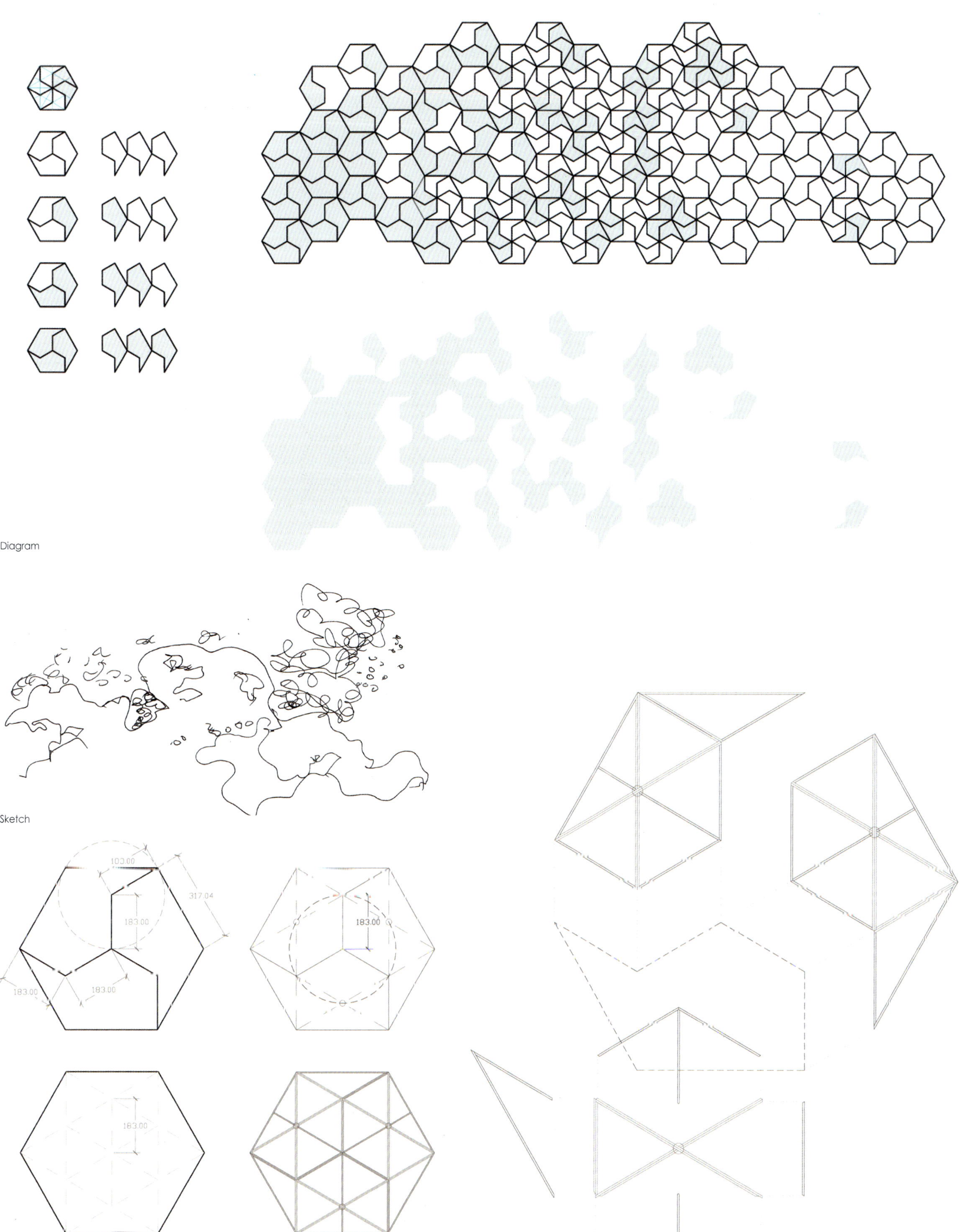

Diagram

Sketch

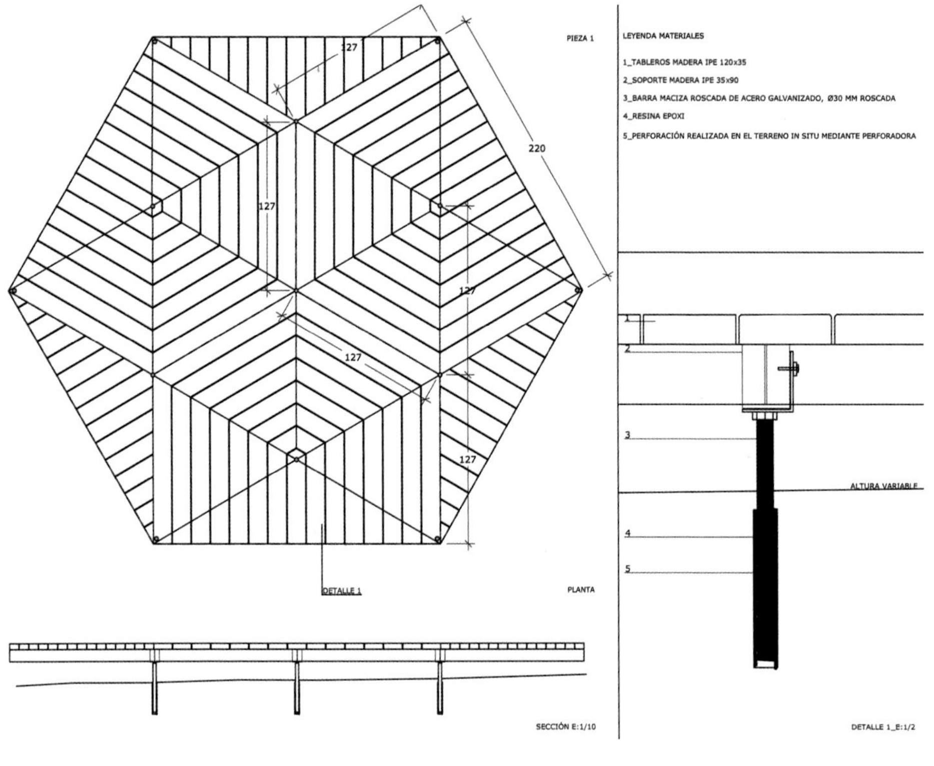
PIEZA 1
LEYENDA MATERIALES
1_TABLEROS MADERA IPE 120x35
2_SOPORTE MADERA IPE 35x90
3_BARRA MACIZA ROSCADA DE ACERO GALVANIZADO, Ø30 MM ROSCADA
4_RESINA EPOXI
5_PERFORACIÓN REALIZADA EN EL TERRENO IN SITU MEDIANTE PERFORADORA
220
127
127
127
127
127
ALTURA VARIABLE
DETALLE 1
PLANTA
SECCIÓN E:1/10
DETALLE 1_E:1/2

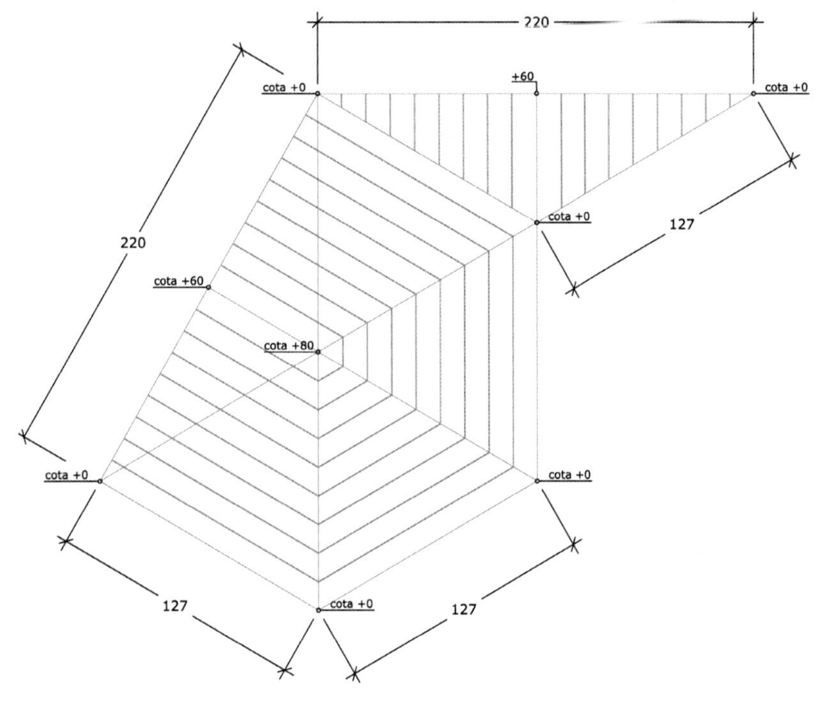
220
+60
cota +0
cota +0
220
cota +0
127
cota +60
cota +80
cota +0
cota +0
127
cota +0
127

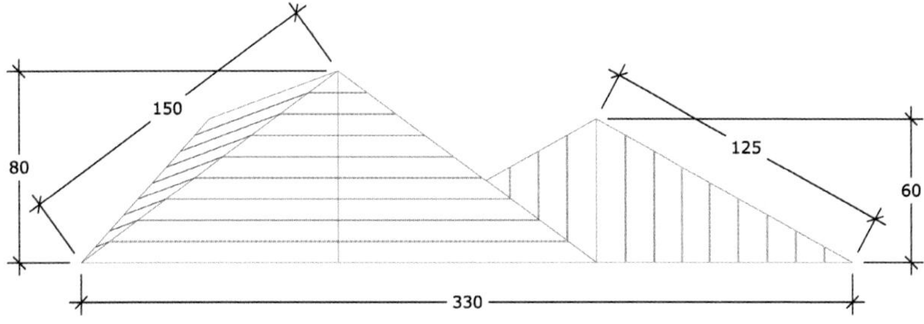
150
80
125
60
330

Detail

© Laura Cantarella, Nuria Díaz

© Laura Cantarella, Nuria Díaz

© Laura Cantarella, Nuria Díaz

Zighizaghi, Favara AG, Italy
OFL Architecture

meri and Riccio Blu

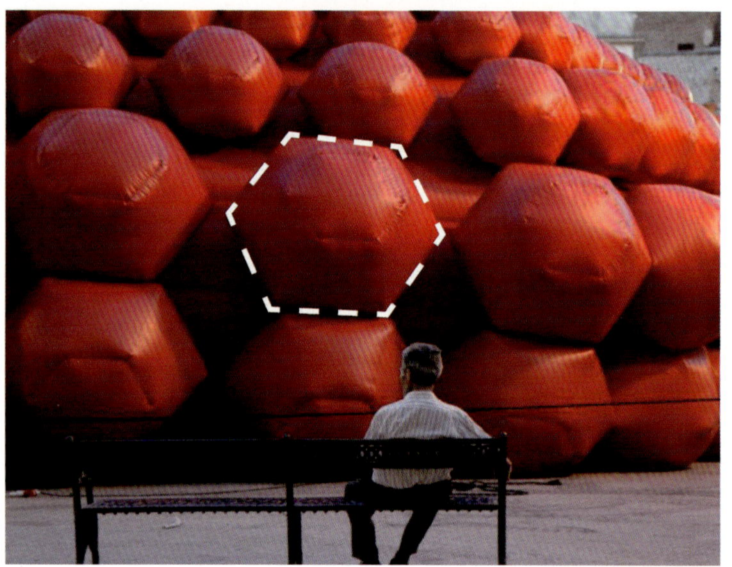

Concept - Hexagon from Pfff

ZIGHI ZAGHI

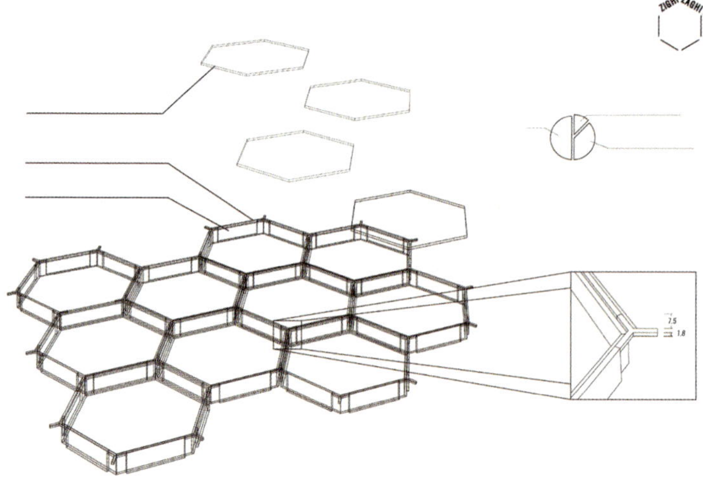

Detail

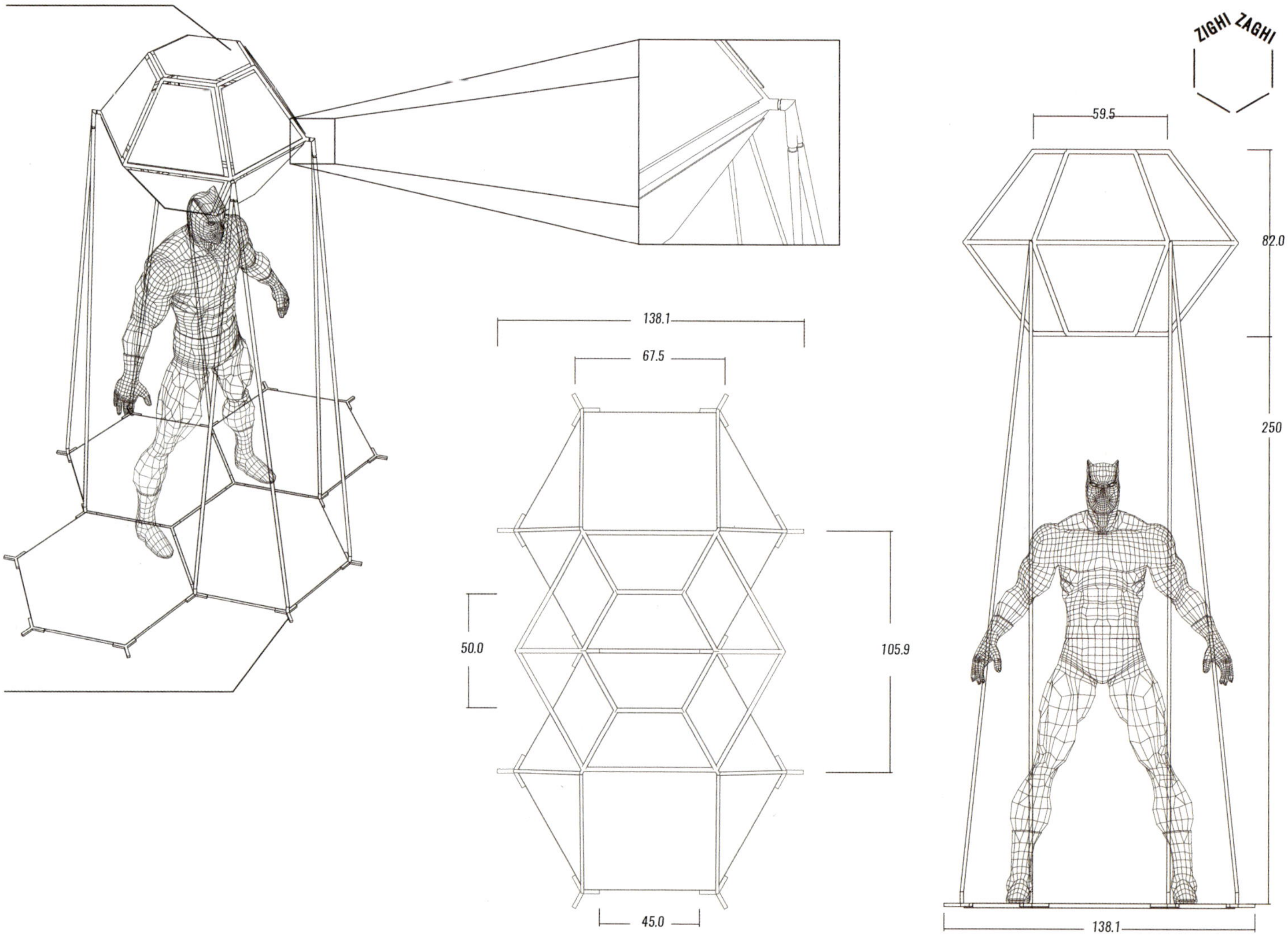

ZIGHI ZAGHI
59.5
82.0
250
138.1
138.1
67.5
50.0
105.9
45.0

Detail

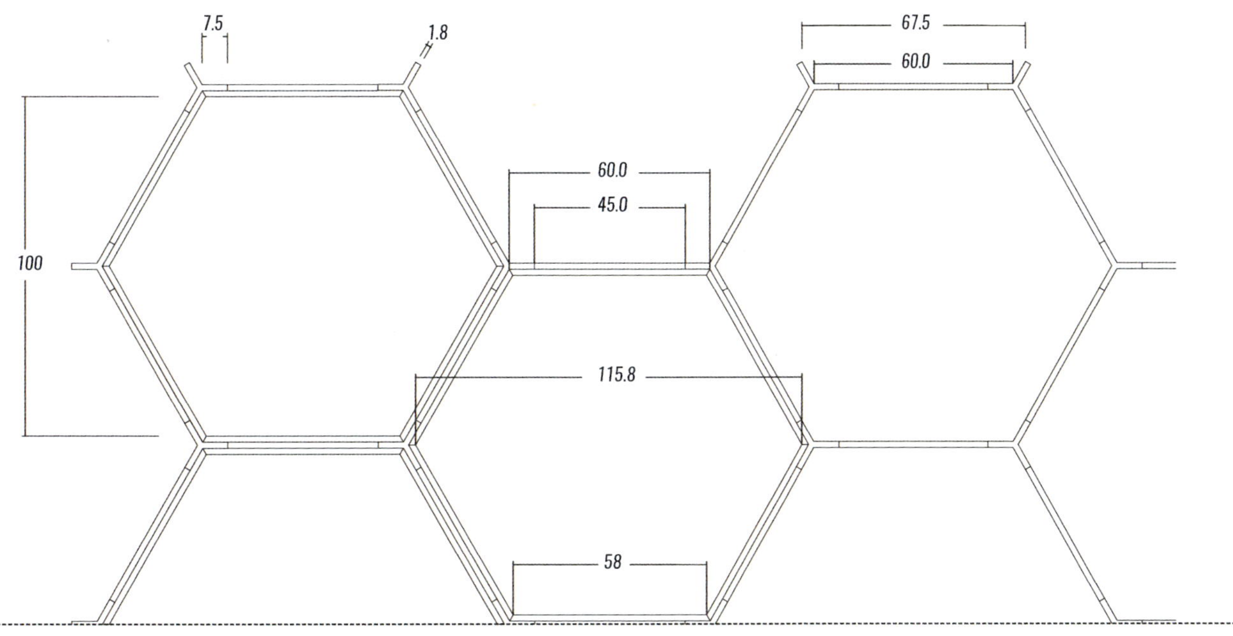

© Riccio Blu

© Riccio Blu

© Guarneri

© Riccio Blu

iamo
tati
apiti
ll'arte"
w.miliashop.com
MILIA
MILIA
MILIA
MILIA
MILIA

© Guarneri

Architectural Material Series

To be Continued